Lecture Notes in Computer Science 12965

More information about this subseries at http://www.springer.com/series/7412

David Svoboda · Ninon Burgos ·
Jelmer M. Wolterink · Can Zhao (Eds.)

Simulation and Synthesis in Medical Imaging

6th International Workshop, SASHIMI 2021 Held in Conjunction with MICCAI 2021 Strasbourg, France, September 27, 2021 Proceedings

Editors
David Svoboda
Masaryk University
Brno, Czech Republic

Ninon Burgos
CNRS – Paris Brain Institute
Paris, France

Jelmer M. Wolterink
University of Twente
Enschede, The Netherlands

Can Zhao
NVIDIA
Santa Clara, CA, USA

ISSN 0302-9743 ISSN 1611-3349 (electronic)
Lecture Notes in Computer Science
ISBN 978-3-030-87591-6 ISBN 978-3-030-87592-3 (eBook)
https://doi.org/10.1007/978-3-030-87592-3

LNCS Sublibrary: SL6 – Image Processing, Computer Vision, Pattern Recognition, and Graphics

This Springer imprint is published by the registered company Springer Nature Switzerland AG
The registered company address is: Gewerbestrasse 11, 6330 Cham, Switzerland

Preface

The Medical Image Computing and Computer Assisted Intervention (MICCAI) community needs data with known ground truth to develop, evaluate, and validate computerized image analytic tools, as well as to facilitate clinical training. Since synthetic data are ideally suited for this purpose, a full range of models underpinning image simulation and synthesis, also referred to as image translation, cross-modality synthesis, image completion, etc., have been developed over the years: (a) machine and deep learning methods based on generative models; (b) simplified mathematical models to test segmentation, tracking, restoration, and registration algorithms; (c) detailed mechanistic models (top-down), which incorporate priors on the geometry and physics of image acquisition and formation processes; and (d) complex spatio-temporal computational models of anatomical variability, organ physiology, and morphological changes in tissues or disease progression.

The goal of the Simulation and Synthesis in Medical Imaging (SASHIMI) workshop is to bring together all those interested in such problems in order to engage in invigorating research, discuss current approaches, and stimulate new ideas and scientific directions in this field. The objectives are to (a) bring together experts on image synthesis to raise the state of the art; (b) hear from invited speakers outside of the MICCAI community, for example in the areas of transfer learning, generative adversarial networks, or variational autoencoders, to cross-fertilize these fields; and (c) identify challenges and opportunities for further research. We also want to identify the suitable approaches to evaluate the plausibility of synthetic data and to collect benchmark data that could help with the development of future algorithms.

The 6th SASHIMI workshop was successfully held in conjunction with the virtual 24th International Conference on Medical Image Computing and Computer Assisted Intervention (MICCAI 2021) as a satellite event on September 27, 2021. Submissions were solicited via a call for papers circulated by the MICCAI organizers, via social media, as well as by directly emailing colleagues and experts in the area. Each of the 18 submissions received underwent a double-blind review by at least two members of the Program Committee, consisting of researchers actively contributing in the area. Compared to the 2020 edition, we introduced three paper categories to better reflect the scope of the individual contributions: papers introducing new methods, application-oriented papers, and demos. At the conclusion of the review process, 14 papers were accepted. Overall, the contributions span the following broad categories in alignment with the initial call for papers: methods based on generative models or adversarial learning for MRI/CT/microscopy image synthesis and applications of image synthesis and simulation for data augmentation, image enhancement, or segmentation.

We would like to thank everyone who contributed to this 6th workshop: the members of the Organizing Committee for their assistance, the authors for their contributions, and the members of the Program Committee for their precious review work, promotion of the workshop, and general support. We are particularly grateful to the invited speaker,

Gordon Wetzstein, who kindly shared his expertise and knowledge with the community. We also thank the MICCAI society for the general support.

September 2021

David Svoboda
Ninon Burgos
Jelmer M. Wolterink
Can Zhao

Organization

Program Chairs

David Svoboda	Masaryk University, Czech Republic
Ninon Burgos	CNRS - Paris Brain Institute, France
Can Zhao	NVIDIA, USA
Jelmer M. Wolterink	University of Twente, The Netherlands

Program Committee

Ninon Burgos	CNRS - Paris Brain Institute, France
Aaron Carass	Johns Hopkins University, USA
Blake Dewey	Johns Hopkins University, USA
Florian Dubost	Stanford University, USA
Hamid Fehri	University of Oxford, UK
Martin Maška	Masaryk University, Czech Republic
Anirban Mukhopadhyay	Technische Universität Darmstadt, Germany
Jack Noble	Vanderbilt University, USA
Dzung Pham	Henry Jackson Foundation, USA
David Svoboda	Masaryk University, Czech Republic
François Varray	Creatis, France
Jelmer M. Wolterink	University of Twente, The Netherlands
Can Zhao	NVIDIA, USA
Ting Zhao	HHMI, Janelia Research Campus, USA

Contents

Method-Oriented Papers

Application-Oriented Papers

Method-Oriented Papers

Detail Matters: High-Frequency Content for Realistic Synthetic MRI Generation

Filip Rusak[1,2(✉)], Rodrigo Santa Cruz[2], Elliot Smith[3], Jurgen Fripp[2], Clinton Fookes[1], Pierrick Bourgeat[2], and Andrew Bradley[1]

[1] Queensland University of Technology, Brisbane, QLD, Australia
rus173@csiro.au
[2] CSIRO, Herston, QLD, Australia
[3] Maxwell Plus, Brisbane, QLD, Australia

Abstract. Deep Learning (DL)-based segmentation methods have been quite successful in various medical imaging applications. The main bottleneck of these methods is the scarcity of quality-labelled samples needed for their training. The lack of labelled training data is often addressed by augmentation methods, which aim to synthesise realistic samples with corresponding labels. While the synthesis of realistic samples remains a challenging task, little is known about the impact of fine detail in synthetic data on the performance of DL-based segmentation models. In this work, we investigate whether, and to what extent, the high-frequency (HF) detail in synthetic brain MR images (MRIs) impacts the performance of DL-based segmentation methods. To assess the impact of HF detail, we generate two synthetic datasets, with and without HF detail and train corresponding segmentation models to evaluate the impact on their performance. The results obtained demonstrate that the presence of HF detail in synthetic brain MRIs, used during training, significantly improve the Dice score up to 1.73% for Gray Matter (GM), 1.34% for White Matter (WM) and 4.41% for Cerebrospinal Fluid (CSF); and therefore justify the need for synthesising realistic-looking MRIs.

Keywords: Data augmentation · Brain MRI · Generative adversarial network · Realistic brain MRI synthesis

1 Introduction

In recent years, Convolutional Neural Networks (CNNs) have been successfully applied on segmentation tasks in the field of medical imaging [5,6,9]. The quality of segmentation results largely depends on the training method, data and, in some cases, corresponding labels required for the segmentation model training [19]. Supervised segmentation methods perform better in comparison to the other

This work was funded in part through an Australian Department of Industry, Energy and Resources CRC-P project between CSIRO, Maxwell Plus and I-Med Radiology Network.

D. Svoboda et al. (Eds.): SASHIMI 2021, LNCS 12965, pp. 3–13, 2021.
https://doi.org/10.1007/978-3-030-87592-3_1

types of learning, but rely on large amounts of quality labelled data which is often not available [16,30]. The reason for a large train set requirement is to provide CNNs with diverse samples, so they can generalise and perform segmentation tasks on unseen samples [10].

Data augmentation is commonly used to bridge the gap between the available and desired quantity of samples, which is reflected as an improvement in segmentation results [7,30]. Data augmentation methods can be broadly grouped as traditional, mixing or synthetic [22]. Traditional augmentation methods such as scaling, rotation, flipping and translation provide limited variation improvements. Mixup [29] in recent years, due to its simplicity, gained popularity in the context of data augmentation for classification, but it has limited applicability on segmentation tasks with sometimes mixed results [4]. Synthetic augmentation methods, for uni or cross-domain scenarios, are commonly realised in some form of Generative Adversarial Networks (GANs) [8]. GANs are capable of producing more diverse samples than traditional augmentation methods, resulting in the improvement of model generalisation ability [28]. Further, conditional GANs [17] and their variations [11,15,31] provide more control over data synthesis and, in the context of brain MR imaging, allow the generation of brain MRIs with particular anatomy, pathology or modality.

Recent work [32] addressed the problem of unrealistic-looking synthetic brain MRI generation (commonly over-smoothed without fine detail) and explained it as a manifestation of distribution mismatch between real and synthetic scans. Intuitively, such an argument motivates the generation of realistic brain MRI samples which remains a challenging task. Challenges in the generation of realistic brain MRI scans come from the versatile nature of MRI representations, 3D shape and the scan sizes [13,23]. The major technical hurdle related to realistic synthetic brain MRIs generation is the limited GPU memory [21,23]. Such an obstacle was previously tackled by working with slices or patches [23], employing series of GANs [21,23] or splitting the generation of shape and texture/appearance [14,30] into separate tasks. The aforementioned methods obtained good results, but often use complex architectures, require training of multiple models or lack control over generated samples. Further, while state-of-the-art methods have focused on the synthesis of realistic medical images [14,21,23,32], little is known about the effect of synthetic image realism on the MRI brain segmentation models.

In this paper, we evaluate the effect of high-frequency (HF) detail, as a single aspect of synthetic brain MRI realism, on the performance of brain segmentation methods. First, we propose a single GAN-based method to generate 3D synthetic MRIs with HF detail. Second, we generate two synthetic brain MRI datasets, with and without HF detail. Then, we evaluate the impact of HF detail on segmentation performance, by training segmentation models on three datasets: i) real MRIs, ii) real + synthetic MRIs with and iii) without HF detail. The obtained results suggest that synthetic can replace real MRIs for the purpose of model training while achieving comparable results. Finally, we conclude that

HF detail matters for the training of brain segmentation models and therefore justify the efforts needed to generate synthetic brain MRIs with HF detail.

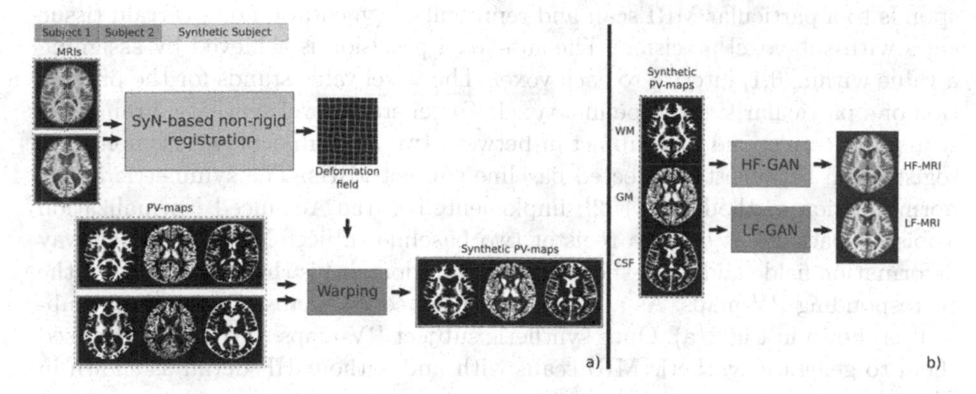

Fig. 1. Non-rigid registration-based MRI and PV-map augmentation (a), LF and HF synthetic brain MRI generation (b).

2 Methods

Data and Pre-processing. 1000 T1-w brain MRI scans from ADNI[1] [12, 27] were used to train, validate and test, both, brain MRI synthesis and segmentation models. For the purpose of training, validation and testing of the GAN model, the 1000 MRI scans were split randomly in the ratio 60:20:20, respectively, without any overlaps between the sets, while retaining the equal distribution of gender and diagnosis. The identical validation and test sets were used for GAN and segmentation models. In the case of segmentation models, the train set was created from 250/600 randomly selected scans included in the GAN models train set. All scans were pre-processed using bias field correction in the brain region of interest (ROI) [24], rigid registration to the MNI-space ($181 \times 217 \times 181$ voxels) and z-score intensity normalisation with the mean value computed from brain ROI. The corresponding labels were derived by a segmentation method implemented based on the expectation-maximisation algorithm [25]. Further, Partial Volume (PV)-maps were estimated from derived labels following the PV-estimation algorithm described in [1].

[1] Data used in the preparation of this article were obtained from the Alzheimer's Disease Neuroimaging Initiative (ADNI) database (adni.loni.usc.edu). The ADNI was launched in 2003 as a public-private partnership, led by Principal Investigator Michael W. Weiner,MD. The primary goal of ADNI has been to test whether serial magnetic resonance imaging (MRI), positron emission tomography (PET), other biological markers, and clinical and neuropsychological assessment can be combined to measure the progression of mild cognitive impairment (MCI) and early Alzheimer's disease (AD). For up-to-date information, see www.adni-info.org.

Brain MRI Synthesis. In this section, we focus on the generation of synthetic MRI scans from three tissue classes represented with PV-maps: WM, GM and CSF. We define a PV-map (M_{pv}) as a volume $M_{pv} \in [0,1]^{w \times h \times d}$ that corresponds to a particular MRI scan and represents segmentation of a certain tissue class with sub-voxel precision. The sub-voxel precision is achieved by assigning a value within [0,1] interval to each voxel. The voxel value stands for the proportion of a particular tissue type in a voxel. To generate novel synthetic brain MRI scans that correspond to a subject in-between two real subjects, we use non-rigid registration between two selected baseline subject MRIs. The symmetric image normalization method (SyN) [2], implemented in the Advanced Normalization Tools [3] package, is used to register two baseline subject MRIs. The half-way deformation field (middle of the normalisation domain) is then used to warp the corresponding PV-maps. As a result, we obtained PV-maps of a synthetic subject as shown in Fig. 1(a). Once synthetic subject PV-maps are created, we used them to generate synthetic MRI scans with and without HF detail as shown in Fig. 1(b).

A conditional GAN-based solution was selected for the generation of more and less realistic synthetic MRIs from PV-maps of three tissues (WM, GM and CSF). We trained the same GAN in two stages, aiming to obtain two models able to generate synthetic brain MRIs of both levels of realism (Fig. 3). For ease of reference, a GAN model capable of brain MRI synthesis without HF detail, from now on will be referred to as LF-GAN. In contrast, a GAN model capable of brain MRI synthesis with HF detail will be referred to as HF-GAN. The model architecture is based on pix2pix framework [11] and is implemented according to the architecture provided in [20] with further adjustments in the second training stage (HF-GAN model). The detailed architecture components used in our implementation is shown in Fig. 2. Both models employed the same U-Net [18] based generator.

The main difference between the first (LF-GAN) and the second (HF-GAN) training stage is in the discriminator being used. In the first training stage, we train a model according to the architecture outlined in [20]. A PatchGAN discriminator, used in [20], is known to be limited in recovering detail on different scales [26], which allows us to obtain synthetic brain MRIs without HF detail. We trained the model, with the stopping criteria defined as generator loss plateaus for at least 10 epochs with fluctuations no more than 0.01. In the second training stage, we resumed the training with the pre-trained generator and the ResNet discriminator, typically used in super-resolution (SR) applications [15]. Training a GAN, with a ResNet discriminator, without pre-training the generator leads to poor results. In SR applications, the input to SR-GANs are low-resolution (LR) images and the outputs are synthetic images with estimated high-resolution detail. In our case, the resolution of the input and output MRIs remains the same but the input MRIs, generated by a pre-trained generator, visually resemble LR images which allow further recovery of HF-detail during the second training stage. The same training criteria was used for both training stages.

To formalise the training of both GAN models, let data of a certain distribution d_x be denoted with x, a generator with G, its output $G(c_{1-3}, z)$ and a discriminator with D. Moreover, we denote three condition variables with c_{1-3}(M_{pv} for three tissue-types) and a noise variable with z. The objective function is defined as follows,

$$\min_{G} \max_{D} \mathbb{E}_{c_{1-3},x}\left[log\left(D\left(c_{1-3},x\right)\right)\right] + \mathbb{E}_{c_{1-3},z}\left[log\left(1 - D\left(c_{1-3}, G\left(c_{1-3},z\right)\right)\right)\right] + \mathbb{E}_{c_{1-3},x,z}\left[\left\|x - G\left(c_{1-3},z\right)\right\|_1\right]. \tag{1}$$

To evaluate the benefit of synthetic HF-MRIs in the context of brain segmentation model training, we trained 15 brain segmentation models, split into three categories according to the train set composition (Real, Real + LF and Real + HF synthetic MRIs). The data split across the models is outlined in Table 2. The primary aim of this experiment is to evaluate the benefits of using HF synthetic brain MRIs for the training of brain segmentation models rather than achieve a Dice score higher than state-of-the-art brain segmentation methods.

GAN Training. Both models were initialised by Adam optimiser and trained, until satisfying the stopping criteria, with the batch size set to one due to the memory limit and the image size. The initial learning rate of 0.0002 was for training both models. The learning rate decay of 2×10^{-6} was introduced in HF-GAN model after 20th epoch.

Segmentation. The segmentation model used for the evaluation of synthesised MRIs is based on vox2vox[2] [5] which was adapted to facilitate the segmentation of four classes: WM, GM, CSF and background. The input to the model is a skull-stripped brain MRI scan and the model generates a binary segmentation map for each of four classes. To fit in the architecture of the segmentation model, the scans and the corresponding binary labels were zero-padded to the size of $256 \times 256 \times 256$ voxels. The binary labels were created from PV-maps by assigning a voxel to the class with the highest PV-value at the same location across three PV-maps (WM, GM and CSF).

Table 1. Image quality assessment by full-reference IQM (left) and segmentation/PV-estimation error of three tissue classes for LF and HF relative to real MRIs (right). The * denotes statistically significant results (t-test after Holm-Bonferroni correction).

Dataset	MS-SSIM ↑	NRMSE ↓	PSNR ↑	NRMSE ↓		
				GM	WM	CSF
LF	0.9834 ± 0.0085	0.0121 ± 0.0423	34.7167 ± 1.9913	0.04343 ± 0.1893	0.0321 ± 0.1465	0.0554 ± 0.1627
HF	**0.9870 ± 0.0071***	**0.0118 ± 0.0437**	**36.5306 ± 1.8113***	**0.0428 ± 0.2806**	**0.0301 ± 0.2852**	**0.0546 ± 0.25092**

[2] The pytorch implementation of vox2vox taken from https://github.com/enochkan/vox2vox.

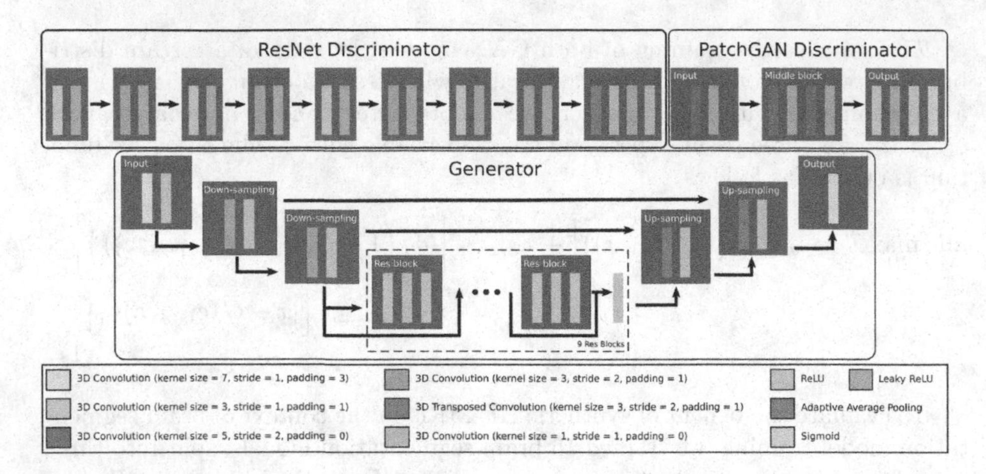

Fig. 2. Architectural building-blocks used for both LF and HF-GANs, and complementary implementation detail.

3 Experiments and Results

Quantitative Evaluation of Synthetic MRIs. To evaluate the synthetic scan quality at both levels of realism (LF vs. HF scans), we reconstructed LF and HF synthetic MRIs from the PV-maps and measured their image quality in relation to the corresponding real MRIs. The PV-maps used for the MRI synthesis were derived from the real MRIs included in the test set (200 samples). Once the synthetic MRIs were generated, the image quality was then measured, between the real and both types of synthetic MRIs, using the following full-reference Image Quality Metrics (IQM): Multiscale Structural Similarity Index Measure (MS-SSIM), Normalised Root Mean Square Error (NRMSE) and Peak Signal-to-Noise Ratio (PSNR). The results, presented in the Table 1 (left), indicate that HF-MRIs have higher similarity with real MRIs compared to the LF-MRI scans. Further, we derived PV-maps from real and synthesised MRIs, and for each tissue class (GM, WM and CSF), we computed the NRMSE between the PV-maps derived from real and both types of synthetic MRIs. The rationale behind that metric suggests that the difference in PV-maps derived from more similar MRIs should be smaller than in the case of the less similar MRIs. The tissue class-wise results, presented in Table 1 (right), indicate a higher similarity between real and HF than between real and LF-MRIs. According to the results presented in the Table 1, the computed IQMs numerically support the higher visual similarity between real and HF synthetic MRIs, compared to the LF synthetic MRIs, as illustrated in Fig. 3.

The Impact of HF Detail on Brain Segmentation. After training the aforementioned segmentation models, we segmented three tissue-classes (GM, WM and CSF) from each of 200 test real MRI samples with each of the trained segmentation models. The model performance was measured using Dice score. The

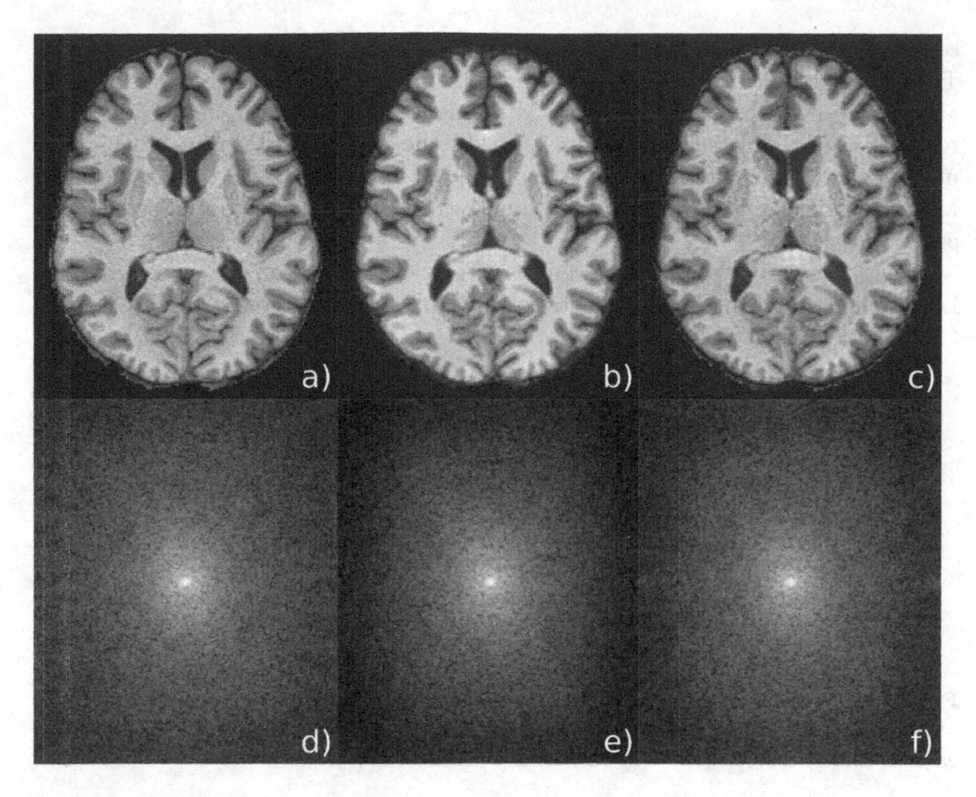

Fig. 3. Real brain MRI (a), synthetic scan generated by LF-GAN (b) and synthetic scan generated by HF-GAN. The respective representation in Fourier domain shown on (d), (e) and (f).

measurement results are presented in the Table 2 and Fig. 4. The upper section of Table 2 shows the segmentation results obtained from five models trained on the real MRIs and corresponding segmentation of the three tissue classes. The bottom section shows the results obtained from the eight remaining segmentation models. According to the results presented in Table 2 and Fig. 4, both LF and HF-MRIs, improve the segmentation model performance for all three tissue classes. It is also evident that the proposed augmentation method can achieve comparable performance to models trained on real MRIs only, with only 20% of real samples. When comparing the overall segmentation performance of models trained on a mixed dataset (real + synthetic MRIs), the models trained on a HF dataset (real + HF synthetic MRIs) perform better than models trained on a LF dataset (real + LF synthetic MRIs). In the case of GM segmentation, the models trained on 50 real and 100 as well as 150 HF synthetic MRIs shows statistically significant improvement in comparison to models trained on the same number of LF synthetic MRI samples. In the case of WM segmentation, models trained on a mixed dataset of 50 real and 50, 100 and 150 HF synthetic MRIs as well as 250 real and 200 HF synthetic show statistically significant improvement in comparison to models trained on the same number of LF synthetic MRI samples. The models trained on 50 real MRIs and 100 and 150 synthetic HF MRIs achieve almost the same performance, in WM segmentation, as the models trained on real MRIs of the same data sample size. In the case of CSF segmentation, all models trained on a HF datasets, except the last model (250 real & 200 synthetic MRIs), show statistically significant improvement comparing to models trained on LF datasets.

The largest Dice score improvement of models trained on HF, compared to, LF dataset, regarding GM segmentation is 1.73%, WM segmentation 1.34%, and CSF segmentation is 4.41%.

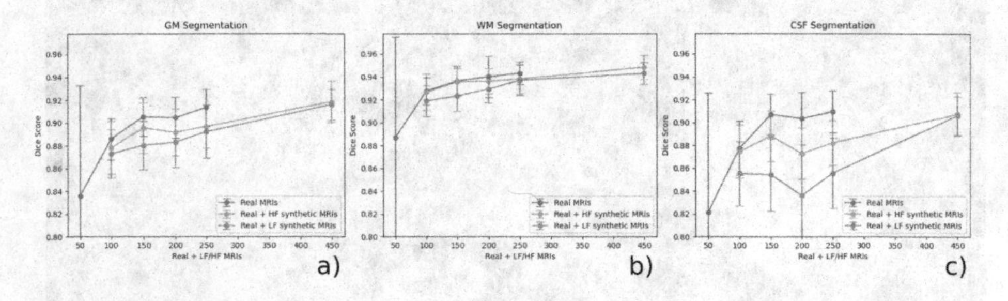

Fig. 4. Performance comparison of segmentation models, represented as a mean Dice score and standard deviation ($\pm\sigma$), trained on real and LF/HF synthetic MRIs, for each of three tissue classes: GM (a), WM (b) and CSF (c).

Table 2. Dice scores of 15 segmentation models trained on datasets with different data splits. The * denotes statistically significant results (t-test after Holm-Bonferroni correction).

Data split (real : synthetic)	GM		WM		CSF	
50:0	0.8356 ± 0.0882		0.8869 ± 0.0967		0.8211 ± 0.1046	
100:0	0.8857 ± 0.018		0.9277 ± 0.0118		0.8775 ± 0.0236	
150:0	0.9051 ± 0.0169		0.9364 ± 0.0132		0.9071 ± 0.0178	
200:0	0.9044 ± 0.0182		0.9401 ± 0.0182		0.9035 ± 0.0228	
250:0	0.9131 ± 0.0161		0.9433 ± 0.01		0.9094 ± 0.0184	
	LF	HF	LF	HF	LF	HF
50:50	0.8727 ± 0.0213	**0.8779 ± 0.0231**	0.919 ± 0.0135	**0.9267 ± 0.016***	0.8554 ± 0.0282	**0.8747 ± 0.0231***
50:100	0.8802 ± 0.0212	**0.8954 ± 0.022***	0.9233 ± 0.0133	**0.9357 ± 0.0116***	0.8544 ± 0.0316	**0.8883 ± 0.0221***
50:150	0.8829 ± 0.0223	**0.8917 ± 0.0208***	0.9294 ± 0.012	**0.9362 ± 0.0111***	0.8361 ± 0.0364	**0.873 ± 0.0224***
50:200	0.8921 ± 0.0229	**0.8966 ± 0.0189**	0.9371 ± 0.0131	**0.9382 ± 0.0129**	0.8557 ± 0.0305	**0.8822 ± 0.0194***
250:200	0.9157 ± 0.0135	**0.9183 ± 0.0180**	0.943 ± 0.0097	**0.9485 ± 0.0106***	0.9062 ± 0.0169	**0.9074 ± 0.0188**

4 Conclusion

In this paper, we proposed a simple, yet effective, GAN training method for the generation of realistic-looking synthetic MRIs and evaluated the impact of

HF detail on the performance of a DL-based brain segmentation method. The results suggest that, in the context of segmentation model training, real data can be replaced by synthetic MRIs while still achieving comparable results. We further showed that synthetic MRIs with HF detail can significantly improve segmentation results on a downstream task, indicating that the realism of synthetic samples matters. For future work, we plan a comprehensive evaluation on the impact of HF detail on other state-of-the-art DL-based brain segmentation methods.

References

1. Acosta, O., et al.: Automated voxel-based 3D cortical thickness measurement in a combined Lagrangian-Eulerian PDE approach using partial volume maps. Med. Image Anal. **13**(5), 730–743 (2009)
2. Avants, B.B., Epstein, C.L., Grossman, M., Gee, J.C.: Symmetric diffeomorphic image registration with cross-correlation: evaluating automated labeling of elderly and neurodegenerative brain. Med. Image Anal. **12**(1), 26–41 (2008)
3. Avants, B.B., Tustison, N., Song, G.: Advanced normalization tools (ANTS). Insight J **2**(365), 1–35 (2009)
4. Chen, C., et al.: Realistic adversarial data augmentation for MR image segmentation. In: Martel, A.L., et al. (eds.) MICCAI 2020. LNCS, vol. 12261, pp. 667–677. Springer, Cham (2020). https://doi.org/10.1007/978-3-030-59710-8_65
5. Cirillo, M.D., Abramian, D., Eklund, A.: Vox2vox: 3D-GAN for brain tumour segmentation. arXiv preprint arXiv:2003.13653 (2020)
6. Coupé, P., et al.: AssemblyNet: a large ensemble of CNNs for 3D whole brain MRI segmentation. Neuroimage **219**, 117026 (2020)
7. Eaton-Rosen, Z., Bragman, F., Ourselin, S., Cardoso, M.J.: Improving data augmentation for medical image segmentation (2018)
8. Goodfellow, I., et al.: Generative adversarial nets. In: Advances in neural information processing systems, pp. 2672–2680 (2014)
9. Henschel, L., Conjeti, S., Estrada, S., Diers, K., Fischl, B., Reuter, M.: FastSurfer-a fast and accurate deep learning based neuroimaging pipeline. NeuroImage 117012 (2020)
10. Huang, S.G., Chung, M.K., Qiu, A., Initiative, A.D.N.: Fast mesh data augmentation via chebyshev polynomial of spectral filtering. arXiv preprint arXiv:2010.02811 (2020)
11. Isola, P., Zhu, J.Y., Zhou, T., Efros, A.A.: Image-to-image translation with conditional adversarial networks. In: Proceedings of the IEEE Conference on Computer Vision and Pattern Recognition, pp. 1125–1134 (2017)
12. Jack Jr., C.R., et al.: The Alzheimer's disease neuroimaging initiative (ADNI): MRI methods. J. Magn. Reson. Imaging: Off. J. Int. Soc. Magn. Reson. Med. **27**(4), 685–691 (2008)
13. Jog, A., Carass, A., Roy, S., Pham, D.L., Prince, J.L.: MR image synthesis by contrast learning on neighborhood ensembles. Med. Image Anal. **24**(1), 63–76 (2015)
14. Keong, C.C., Wei, H.E.T.: Synthesis of 3D MRI brain images with shape and texture generative adversarial deep neural networks. IEEE Access **9**, 64747–64760 (2021)

15. Ledig, C., et al.: Photo-realistic single image super-resolution using a generative adversarial network. In: Proceedings of the IEEE Conference on Computer Vision and Pattern Recognition, pp. 4681–4690 (2017)
16. Lee, J., Kim, E., Lee, S., Lee, J., Yoon, S.: FickleNet: weakly and semi-supervised semantic image segmentation using stochastic inference. In: Proceedings of the IEEE Conference on Computer Vision and Pattern Recognition, pp. 5267–5276 (2019)
17. Mirza, M., Osindero, S.: Conditional generative adversarial nets. arXiv preprint arXiv:1411.1784 (2014)
18. Ronneberger, O., Fischer, P., Brox, T.: U-net: convolutional networks for biomedical image segmentation. In: Navab, N., Hornegger, J., Wells, W.M., Frangi, A.F. (eds.) MICCAI 2015. LNCS, vol. 9351, pp. 234–241. Springer, Cham (2015). https://doi.org/10.1007/978-3-319-24574-4_28
19. Roy, A.G., Conjeti, S., Navab, N., Wachinger, C., Initiative, A.D.N., et al.: QuickNAT: a fully convolutional network for quick and accurate segmentation of neuroanatomy. Neuroimage **186**, 713–727 (2019)
20. Rusak, F., et al.: 3D brain MRI GAN-based synthesis conditioned on partial volume maps. In: Burgos, N., Svoboda, D., Wolterink, J.M., Zhao, C. (eds.) SASHIMI 2020. LNCS, vol. 12417, pp. 11–20. Springer, Cham (2020). https://doi.org/10.1007/978-3-030-59520-3_2
21. Sun, L., Chen, J., Xu, Y., Gong, M., Yu, K., Batmanghelich, K.: Hierarchical amortized training for memory-efficient high resolution 3D GAN. arXiv preprint arXiv:2008.01910 (2020)
22. Tajbakhsh, N., Jeyaseelan, L., Li, Q., Chiang, J.N., Wu, Z., Ding, X.: Embracing imperfect datasets: a review of deep learning solutions for medical image segmentation. Med. Image Anal. **63**, 101693 (2020)
23. Uzunova, H., Ehrhardt, J., Jacob, F., Frydrychowicz, A., Handels, H.: Multi-scale GANs for memory-efficient generation of high resolution medical images. In: Shen, D., et al. (eds.) MICCAI 2019. LNCS, vol. 11769, pp. 112–120. Springer, Cham (2019). https://doi.org/10.1007/978-3-030-32226-7_13
24. Van Leemput, K., Maes, F., Vandermeulen, D., Suetens, P.: Automated model-based bias field correction of MR images of the brain. IEEE Trans. Med. Imaging **18**(10), 885–896 (1999)
25. Van Leemput, K., Maes, F., Vandermeulen, D., Suetens, P.: Automated model-based tissue classification of MR images of the brain. IEEE Trans. Med. Imaging **18**(10), 897–908 (1999)
26. Wang, J., Chen, Y., Wu, Y., Shi, J., Gee, J.: Enhanced generative adversarial network for 3D brain MRI super-resolution. In: Proceedings of the IEEE/CVF Winter Conference on Applications of Computer Vision, pp. 3627–3636 (2020)
27. Weiner, M.W., et al.: The Alzheimer's disease neuroimaging initiative 3: continued innovation for clinical trial improvement. Alzheimer's Dementia **13**(5), 561–571 (2017)
28. Yi, X., Walia, E., Babyn, P.: Generative adversarial network in medical imaging: a review. Med. Image Anal. **58**, 101552 (2019)
29. Zhang, H., Cisse, M., Dauphin, Y.N., Lopez-Paz, D.: mixup: beyond empirical risk minimization. arXiv preprint arXiv:1710.09412 (2017)
30. Zhao, A., Balakrishnan, G., Durand, F., Guttag, J.V., Dalca, A.V.: Data augmentation using learned transformations for one-shot medical image segmentation. In: Proceedings of the IEEE Conference on Computer Vision and Pattern Recognition, pp. 8543–8553 (2019)

31. Zhu, J.Y., Park, T., Isola, P., Efros, A.A.: Unpaired image-to-image translation using cycle-consistent adversarial networks. In: Proceedings of the IEEE International Conference on Computer Vision, pp. 2223–2232 (2017)
32. Zuo, L., et al.: Synthesizing realistic brain MR images with noise control. In: Burgos, N., Svoboda, D., Wolterink, J.M., Zhao, C. (eds.) SASHIMI 2020. LNCS, vol. 12417, pp. 21–31. Springer, Cham (2020). https://doi.org/10.1007/978-3-030-59520-3_3

Joint Image and Label Self-super-Resolution

Samuel W. Remedios[1(✉)], Shuo Han[2], Blake E. Dewey[3], Dzung L. Pham[4], Jerry L. Prince[3], and Aaron Carass[3]

[1] Department of Computer Science, Johns Hopkins University, Baltimore, MD 21218, USA
samuel.remedios@jhu.edu
[2] Department of Biomedical Engineering, Johns Hopkins University, Baltimore, MD 21218, USA
[3] Department of Electrical Engineering, Johns Hopkins University, Baltimore, MD 21218, USA
[4] Center for Neuroscience and Regenerative Medicine, Henry M. Jackson Foundation, Bethesda, MD 20817, USA

Abstract. We propose a method to jointly super-resolve an anisotropic image volume along with its corresponding voxel labels without external training data. Our method is inspired by internally trained super-resolution, or self-super-resolution (SSR) techniques that target anisotropic, low-resolution (LR) magnetic resonance (MR) images. While resulting images from such methods are quite useful, their corresponding LR labels—derived from either automatic algorithms or human raters—are no longer in correspondence with the super-resolved volume. To address this, we develop an SSR deep network that takes both an anisotropic LR MR image and its corresponding LR labels as input and produces both a super-resolved MR image and its super-resolved labels as output. We evaluated our method with 50 T_1-weighted brain MR images 4× down-sampled with 10 automatically generated labels. In comparison to other methods, our method had superior Dice across all labels and competitive metrics on the MR image. Our approach is the first reported method for SSR of paired anisotropic image and label volumes.

Keywords: Super-resolution · MRI · Segmentation

1 Introduction

Volumetric labels serve many purposes in medical imaging but primarily allow for scientific analysis on regions of interest. The assignment of labels to voxels is called segmentation and can be performed by an automatic algorithm or a trained human rater. In both cases, delineations are made on the MR volume in question. Isotropic MR volumes are rarely segmented fully by raters due to the manual labor required, resulting in partially annotated volumes. Furthermore,

D. Svoboda et al. (Eds.): SASHIMI 2021, LNCS 12965, pp. 14–23, 2021.
https://doi.org/10.1007/978-3-030-87592-3_2

anisotropic acquisitions of volumetric medical images are more common in practice for a variety of reasons including preservation of high signal-to-noise ratio for in-plane signals, increased patient comfort, and a reduction in scan time. Manual delineations of such data produce necessarily anisotropic label volumes.

Towards the reduction of human labor, many algorithms have been developed to perform automatic segmentation [8,9,18,20,23]. These methods generally either 1) only accept isotropic volumes, 2) up-sample or super-resolve anisotropic volumes prior to segmentation, or 3) perform 2D segmentations in-plane and stack the resulting slices. Additionally, changes in MR acquisition protocols, as well as anatomical changes (including the presence of pathology), can prevent the use of existing data sets for training machine learning algorithms.

Our goal is to enable the use of retrospective anisotropic labels generated by trained raters or trustworthy algorithms when appropriate external training data is unavailable. Via our proposed joint image and label super-resolution (SR), we aim to address three concerns: the burden of human raters to fully segment isotropic volumes, the reliance on external interpolation techniques which may not be designed for discrete labels, and the instability of existing downstream methods which require isotropic data to be used on retrospectively accessed anisotropic data. To demonstrate this contribution, we applied our method to simulated low resolution (LR) images and multi-class labels and evaluated the results both qualitatively and quantitatively. We compared to three methods and found that our method produced superior labels to the others and simultaneously achieved competitive MR SR results.

1.1 Related Work

Self-super-resolution. Supervised SR deep networks have been applied to knee [2], brain [3,7,17], and heart [16] MRI to produce high resolution (HR) volumes. However, due to the lack of paired LR-HR data, self-super-resolution (SSR) [10] was proposed to super-resolve a volume without external training data. These methods take advantage of the inherent internal training data in anisotropic volumes. By convention, anisotropic MR images are acquired at HR in-plane and LR through-plane and the 3D volume can be examined in 2D along the three cardinal axes. This framing allows for the in-plane slices themselves to be a type of HR ground truth (although still blurred through-plane), and both realizations of through-plane slices to be the test-time LR input data for an algorithm. In the SSR framework, the in-plane HR slices undergo HR $\rightarrow$ LR degradation to create simulated training pairs for supervised learning. After the model is fit, it is applied to both 2D through-plane realizations and aggregated. A deep network variant was introduced for MR SSR by Zhao et al. [22].

Label Super-Resolution. Despite the many advances in SR for MRI, there have been few works to increase the resolution of the corresponding label volumes. Only recently Di et al. [14] have tackled this problem directly, proposing a SR $\rightarrow$ segmentation $\rightarrow$ correction framework. While this method achieves HR labels, it currently is implemented only for binary masks, operates in 2D, and

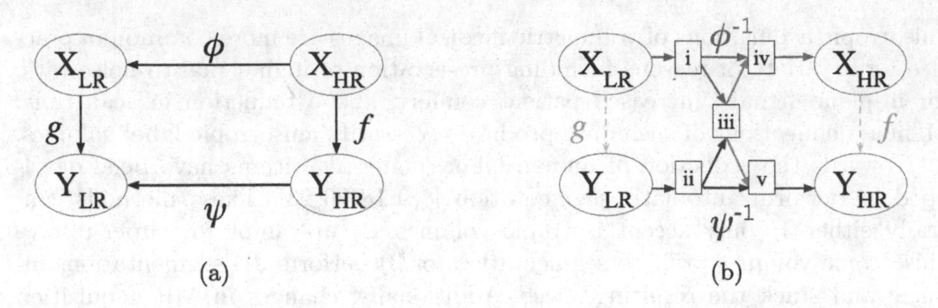

Fig. 1. **(a)** Our observation model assumes the depicted relationship between $X_{\mathrm{LR}}, X_{\mathrm{HR}}, Y_{\mathrm{LR}}$, and Y_{HR}. **(b)** We aim to learn the inverse problems ϕ^{-1} and ψ^{-1}. Nodes labeled $i - v$ are learnable blocks in a deep CNN, shown in Fig. 2.

relies on an external SR algorithm and iterative improvement of a segmentation; consequently the approach does not allow shared information to flow between the LR MR and LR labels during the training process.

Delannoy et al. [5] proposed SegSRGAN that first super-resolves the MRI and then segments the resulting HR volume. At train time their method operates on 3D data, relies on an external dataset consisting of paired HR MR and label images, creates LR MR training data via simulation, does not make use of existing LR labels, and in the current implementation only operates on binary classes.

Our Contribution. Our proposed method[2] is the first to super-resolve multi-class label volumes and is also the first to jointly super-resolve an MR volume alongside its label volume. We accomplish this without external training data using the aforementioned SSR framework: a 10-layer dual input-output convolutional neural network (CNN) with a shared processing branch and individual long skip-connections, a sampling and class weighting strategy to permit the model to learn evenly across all label and non-label regions, and a compound loss function which balances both MR and label error during training. Our model creates internal training data, fits an anti-aliasing and SR network to this data, and produces both HR volumes in under 6 minutes total on a Tesla V100.

2 Methods

2.1 Formulation

Let the underlying HR image volume be X_{HR} and its corresponding label volume be Y_{HR}. We assume there is a direct relation f such that $f(X_{\mathrm{HR}}) = Y_{\mathrm{HR}}$. Similarly, let the acquired LR image volume be X_{LR} and its corresponding label volume be Y_{LR}, with $g(X_{\mathrm{LR}}) = Y_{\mathrm{LR}}$. We assume that we can obtain X_{LR} from X_{HR} as $\phi(X_{\mathrm{HR}}) = X_{\mathrm{LR}}$ and accordingly define $\psi(Y_{\mathrm{HR}}) = Y_{\mathrm{LR}}$; these relations are illustrated in Fig. 1(a).

[2] Source code is publicly available at https://gitlab.com/iacl/label_smore.

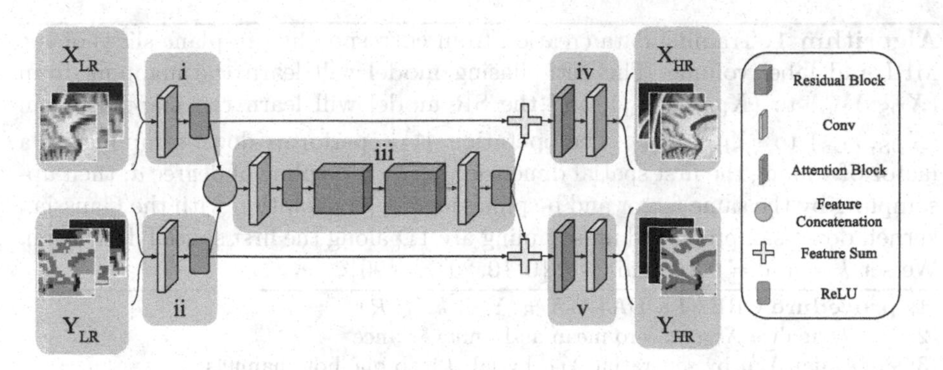

Fig. 2. The joint SR architecture expects a digitally up-sampled spatially LR pair of images and produces a spatially SR pair of images. The annotated blocks correspond to Fig. 1(b). Residual blocks (with $0.1\times$ scaling) are identical to EDSR [13]. Attention blocks ($r = 1$) are identical to RCAN [21].

For the joint image-and-label SR task, we aim to model the inverse functions ϕ^{-1} and ψ^{-1}; this is shown in Fig. 1(b). However, ϕ^{-1} and ψ^{-1} are ill-posed as there are multiple HR solutions for a given LR input. In light of this, we introduce a shared processing path, in our CNN depicted in Fig. 2.

Information from both X_{LR} and Y_{LR} flow through *iii* and encourage joint feature extraction. Thus, the paths $i \rightarrow iv$ and $i \rightarrow iii \rightarrow iv$ model ϕ^{-1} and the paths $ii \rightarrow v$ and $ii \rightarrow iii \rightarrow v$ model ψ^{-1}. Further, while our method is internally trained and creates simulated LR data, it is still supervised learning and gives rise to two scenarios. First, since patches contain spatial context, one-to-one mappings from LR $\rightarrow$ HR may arise; in these scenarios the network can approximate the solution directly. Second, for mappings that are many-to-one, since we train the network in batches with a variant of the L_1 loss with a mean reduction, the model approximates (at best) the mean of all solutions [12].

2.2 Label Creation

To evaluate our method, we used the OASIS3 dataset [11] and randomly selected 50 T_1-weighted images. We simulated LR acquisition by blurring and down-sampling all MR volumes by a factor of 4. Since ground truth labels are needed for quantitative evaluation, we simulated a human rater using SLANT [9] to automatically and independently segment each of the LR and HR volumes. For each volume, we mapped the 133 labels from SLANT to ten labels: background, ventricles, cerebellar GM, cerebellar WM, cerebral GM, cerebral WM, caudate, putamen, thalamus, and brainstem.

2.3 Training Data Creation

Since our method is internally trained, we create training data pairs by degrading HR in-plane patches present in the input volumes as in [22]. We assume the

Algorithm 1. Training data creation from corresponding in-plane slices of the MRI and label volume. The anti-aliasing model will learn the mapping from $(X_{\text{DS}}, Y_{\text{DS}})$ to $(X_{\text{Blur}}, Y_{\text{Blur}})$, and the SR model will learn the mapping from $(X_{\text{DS}}, Y_{\text{DS}})$ to $(X_{\text{HR}}, Y_{\text{HR}})$. The operation $\downarrow\uparrow_z^{(d)}$ performs down-sampling by a factor of z along the first spatial dimension using a b-spline of degree d, then up-sampling by the same factor and b-spline degree. Convolution with the Gaussian kernel, down-sampling, and up-sampling are 1D along the first spatial dimension. We set $k = 4$, $a = 0.2k$, and $\mathcal{R} = \{0, 10, 20, \ldots, 90\}$.

1: **procedure** CREATE-DATA(X_{LR}, Y_{LR}, k, a, $\mathcal{R}$)
2: Normalize X_{LR} to zero mean and unit variance
3: Assign $\widetilde{Y}_{\text{LR}}$ by separating Y_{LR} by label into one-hot channels
4: Initialize dataset lists $\mathcal{D}$, $\mathcal{D}_{\text{Blur}}$, $\mathcal{D}_{\text{DS}}$
5: **for** $r \in \mathcal{R}$ **do**
6: Sample both k'_{Blur} & k'_{DS} from $\mathcal{U}(k-a, k+a)$
7: Construct a 1D Gaussian kernel s with FWHM k'_{Blur}
8: Rotate 2D image & labels: X_{LR} & $\widetilde{Y}_{\text{LR}}$ in-plane by r
9: Apply blur: $X_{\text{Blur}} \leftarrow X_{\text{LR}} * s$ & $\widetilde{Y}_{\text{Blur}} \leftarrow \widetilde{Y}_{\text{LR}} * s$
10: Down-sample and up-sample: $X_{\text{DS}} \leftarrow X_{\text{Blur}} \downarrow\uparrow_{k'_{\text{DS}}}^{(1)}$ & $\widetilde{Y}_{\text{DS}} \leftarrow \widetilde{Y}_{\text{Blur}} \downarrow\uparrow_{k'_{\text{DS}}}^{(0)}$
11: $Y_{\text{Blur}} \leftarrow$ channel-wise argmax $\widetilde{Y}_{\text{Blur}}$
12: $Y_{\text{DS}} \leftarrow$ channel-wise argmax $\widetilde{Y}_{\text{DS}}$
13: Append $(X_{\text{LR}}, Y_{\text{LR}})$ to $\mathcal{D}$, $(X_{\text{Blur}}, Y_{\text{Blur}})$ to $\mathcal{D}_{\text{Blur}}$, $(X_{\text{DS}}, Y_{\text{DS}})$ to $\mathcal{D}_{\text{DS}}$
14: **end for**
15: **end procedure**

blur kernel for the MRI is Gaussian with full-width-at-half-maximum (FWHM) equal to the slice thickness. We convolve this kernel along one axis to blur the data in 1D and, since this kernel is not a sufficient low-pass-filter, we introduce aliasing via linearly interpolated down-sampling. We repeat the same process for the label volume, except the down-sampling instead uses nearest neighbor interpolation. Additionally, prior to blur kernel convolution and down-sampling, we represent the labels as a one-hot encoded volume to prevent semantic leakage when degrading labels. Finally, we resize the blurred and down-sampled images to the same digital grid size by up-sampling with the corresponding interpolation: linear for the MR volume and nearest neighbor for the label volume. This allows us to include differing degradation factors in the same batch during training.

We define the ratio between in-plane and through-plane to be k (in our experiments, $k = 4$). We also define a and sample the blur kernel FWHM and down-sample ratio independently from the uniform distribution $\mathcal{U}(k-a, k+a)$ (as in [4]), with $a = 0.2k$ due to the leniency in reported resolution on *calibrated* scanners of 20% [1]. This has an additional benefit beyond accounting for image header inaccuracies; when the blur kernel FWHM is less than the slice separation, a slice gap occurs (and the converse implies slice overlap). By sampling different blur kernel FWHMs and down-sample factors, we allow our model to "see" data from both of these scenarios. To expand the training data, we additionally apply

2D in-plane rotations to the volume as data augmentation, and train the model by randomly sampling 64×64 pixel patches. See Algorithm 1 for more details.

2.4 Implementation

We implemented our network in PyTorch and trained the architecture from Fig. 2 end-to-end after creating training data, see Sect. 2.3. Note that the architecture shown in Fig. 2 correctly depicts the number of layers used in our method. The residual blocks [13] and attention blocks [21] each consist of two convolutional layers. All convolutional layers have 256 filters; thus, our network is 10 layers deep with approximately 4 million parameters.

We train for 3,750 steps with a batch size of 8 using automatic mixed precision, with the `AdamW` [15] optimizer, and `OneCycleLR` [19] as the learning rate scheduler with a peak of 1×10^{-4}. Our loss function consists of terms from the image and label branches: $\mathscr{L} = \lambda_1 \mathscr{L}_{\text{image}} + \lambda_2 \mathscr{L}_{\text{label}}$, with $\lambda_1 = 10$ and $\lambda_2 = 1$, chosen empirically. For the image branch, $\mathscr{L}_{\text{image}}$ is the evenly-weighted sum of L_1 and Sobel edge losses, as in [22]. For the label branch, we use cross-entropy loss with class weights. Defining the number of voxels for label i as v_i, with $i = 1, 2, \ldots, C$ and C the total number of non-background label classes, we calculate the class weights w_i for the ith class as $w_i = \log((\max v)/v_i)$, clamped below to a minimum weight of 1.

During training, 64×64 2D in-plane patches are extracted randomly and fed into the network. Patches are sampled such that smaller volume labels are presented to the model more frequently than larger volume labels; the probability for a patch containing label i to be selected is $v_i / \sum_j^{C+1} v_j$, where the sum includes the background label class.

As in [22], we fit two independent models with the same architecture to learn an anti-aliasing (AA) and a SR task. The first model learns the mapping from $(X_{\text{DS}}, Y_{\text{DS}}) \rightarrow (X_{\text{Blur}}, Y_{\text{Blur}})$ and the second learns $(X_{\text{DS}}, Y_{\text{DS}}) \rightarrow (X_{\text{HR}}, Y_{\text{HR}})$ as defined in Algorithm 1. Both models are trained sequentially and independently. At test-time, we follow [22] and apply the AA model to one through-plane axis and then apply the SR model along the resulting volume to an orthogonal through-plane direction. We repeat this for the alternate orthogonal through-plane and, for both the MR and label volumes, fuse the results with Fourier Burst Accumulation [6] with the parameter $p = 0$.

2.5 Comparison Methods

We compare our proposed method to other methods which also do not rely on external training data: nearest neighbor interpolation, bicubic b-spline interpolation, and SMORE [22]. Both interpolation methods are common in medical imaging but do not increase high-frequency information. SMORE is an SSR deep learning technique that accepts a single input and is based on EDSR, a much deeper (66 layers vs our 10) and more highly parameterized (approximately 40 million parameters vs our 4 million) network. Notably, all three comparison methods up-sample or super-resolve the MR and label volumes independently; no joint information is used.

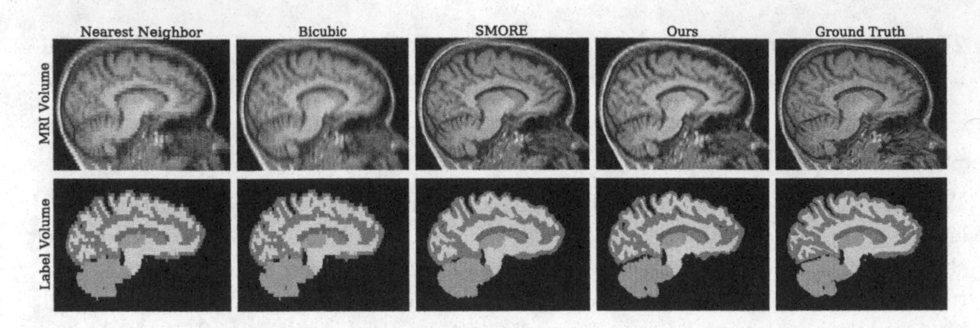

Fig. 3. A representative subject, sagittal view. The columns refer to the method that produced the corresponding volume. In the MR volume, note the aliasing at the top of the skull in our method compared to SMORE. Conversely, in the label volume, our method is able to recover more high-frequency detail in the cerebellar region, and is more consistent across the cortical folds and ventricles.

Since there are multiple labels, we separate each label into its own channel, representing the volume in a 4D one-hot encoding. Both nearest neighbor and bicubic b-spline up-sample this volume channel-wise before an argmax returns the volume to a single-channel multi-label form. To super-resolve the label volume, we modified SMORE to take a multi-channel volume as input and train the model with cross-entropy loss.

3 Results

3.1 Qualitative Results

Figure 3 shows the four methods compared to the original HR image (ground truth). Both the nearest neighbor and bicubic interpolated MR volumes exhibit blurring and a lack of high-frequency detail that are present in SMORE and our proposed method. We note that our proposed joint SR method does not fully remove the aliasing artifacts present at the top of the skull, whereas SMORE correctly removes them.

Regarding the labels, both nearest neighbor and bicubic interpolation retain a "blocky" appearance, indicative of their inability to automatically "round out" the corners of the labels. Both SMORE and the proposed method better resolve this blockiness issue, but our proposed method is able to leverage information present in the MR volume to recover finer details which SMORE cannot (see the cerebellar region in Fig. 3).

3.2 Quantitative Results

We used three metrics to quantify the differences across methods: the Dice similarity coefficient (Dice), the structural similarity index measure (SSIM), and peak signal-to-noise ratio (PSNR). Dice was calculated independently for each

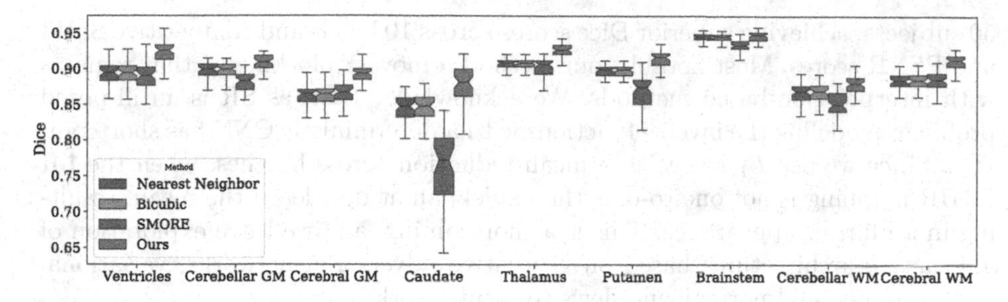

Fig. 4. Volumetric dice scores across 50 subjects by region. Our proposed method significantly outperforms the others across all regions, with $p < 0.05$ as calculated by the sign test.

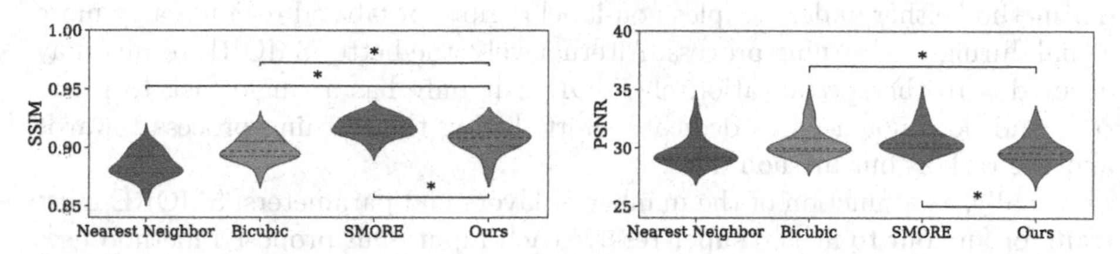

Fig. 5. SSIM and PSNR across 50 subjects. Brackets with asterisks indicate method pairs with significant differences.

of the nine non-background labels and is depicted in Fig. 4. Our proposed method outperformed all others on each structure. While our method has larger variation than interpolation on some smaller structures such as the caudate or putamen, neither nearest neighbor nor bicubic interpolation were able to address the aforementioned qualitative blockiness. To calculate statistical significance, we performed pairwise sign tests on each region between our proposed method and each comparison method, rejecting the null hypothesis that there is a zero-median difference between the calculated Dice values of each method with $p < 0.05$ for all structures.

SSIM and PSNR, illustrated in Fig. 5, were both calculated on the MR volume. We particularly note our method does not outperform SMORE, which corroborates the qualitative results regarding aliasing artifacts. By the sign test, we found our method to achieve significantly higher SSIM than nearest neighbor and bicubic interpolation, but significantly lower SSIM than SMORE. With regards to PSNR, our method does not significantly outperform nearest neighbor interpolation ($p = 0.06$) and is significantly outperformed by both bicubic interpolation and SMORE.

4 Discussion and Conclusion

We have presented a method for joint SSR for anisotropic image and label volumes without the need for external training data. We evaluated our method on

50 subjects, achieving superior Dice scores across 10 labels and competitive SSIM and PSNR scores. Most notably, our method removes "blockiness" that remains with interpolation-based methods. We acknowledge that, as SR is an ill-posed problem, modeling the inverse functions with a deterministic CNN has shortcomings. Since we use L_1 loss with a mean reduction across batches, when the LR $\rightarrow$ HR mapping is not one-to-one, the model can at best learn the mean, resulting in a blurrier appearance. This is a shortcoming, and we leave exploration of alternative architectures based on generative adversarial networks, variational autoencoders, and normalizing flows to future work.

Regarding the difference in MR quantitative metrics, we noted that aliasing was not fully removed with our method as compared to SMORE, especially in the superior skull region. It is possible that without labels for the skull region, our method either under-samples non-label regions or labeled regions carry more signal during the learning process. Alternatively, the better SMORE results may occur due to the specialization of SMORE; it only has a single task to learn, ϕ^{-1}, and does not need to dedicate effort during the learning process towards another task as our method does.

Finally, as a function of the number of layers and parameters, SMORE must train for an hour to fit and super-resolve each input. Our proposed method uses a 10 layer architecture with one-tenth the total trainable parameters, and subsequently produces super-resolved volumes for both the MR and label volumes in under 6 min combined.

Acknowledgements. This material is supported by the National Science Foundation Graduate Research Fellowship under Grant No. DGE-1746891. This work was partially supported by National Multiple Sclerosis Society grant RG-1907-34570, and by the Center for Neuroscience and Regenerative Medicine in the Department of Defense. Theoretical development partially supported by DoD/CDMRP grant MS190131.

References

1. American College of Radiology: American College of Radiology magnetic resonance imaging accreditation program: Phantom test guidance for use of the large MRI phantom for the ACR MRI accreditation program, p. 16 (2018)
2. Chaudhari, A.S., et al.: Super-resolution musculoskeletal MRI using deep learning. Magn. Reson. Med. **80**(5), 2139–2154 (2018)
3. Chen, Y., Xie, Y., Zhou, Z., Shi, F., Christodoulou, A.G., Li, D.: Brain MRI super resolution using 3D deep densely connected neural networks. In: 2018 IEEE 15th International Symposium on Biomedical Imaging (ISBI 2018), pp. 739–742 (2018)
4. Chen, Y., Liu, S., Wang, X.: Learning continuous image representation with local implicit image function. arXiv preprint arXiv:2012.09161 (2020)
5. Delannoy, Q., et al.: SegSRGAN: super-resolution and segmentation using generative adversarial networks - application to neonatal brain MRI. Computers in Biology and Medicine **120**, 103755 (2020)
6. Delbracio, M., Sapiro, G.: Removing camera shake via weighted Fourier burst accumulation. IEEE Trans. Image Process. **24**(11), 3293–3307 (2015)

7. Du, J., et al.: Super-resolution reconstruction of single anisotropic 3D MR images using residual convolutional neural network. Neurocomputing **392**, 209–220 (2020)
8. He, Y., et al.: Structured layer surface segmentation for retina OCT using fully convolutional regression networks. Med. Image Anal. **68**, 101856 (2021)
9. Huo, Y., et al.: 3D whole brain segmentation using spatially localized atlas network tiles. Neuroimage **194**, 105–119 (2019)
10. Jog, A., Carass, A., Prince, J.L.: Self super-resolution for magnetic resonance images. In: Ourselin, S., Joskowicz, L., Sabuncu, M.R., Unal, G., Wells, W. (eds.) MICCAI 2016. LNCS, vol. 9902, pp. 553–560. Springer, Cham (2016). https://doi.org/10.1007/978-3-319-46726-9_64
11. LaMontagne, P.J., et al.: OASIS-3: longitudinal neuroimaging, clinical, and cognitive dataset for normal aging and Alzheimer disease. medRxiv (2019)
12. Ledig, C., et al.: Photo-realistic single image super-resolution using a generative adversarial network. In: Proceedings of the IEEE Conference on Computer Vision and Pattern Recognition, pp. 4681–4690 (2017)
13. Lim, B., Son, S., Kim, H., Nah, S., Mu Lee, K.: Enhanced deep residual networks for single image super-resolution. In: Proceedings of the IEEE Conference on Computer Vision and Pattern Recognition (CVPR) Workshops, July 2017
14. Liu, D., Liu, J., Liu, Y., Tao, R., Prince, J.L., Carass, A.: Label super resolution for 3D magnetic resonance images using deformable U-net. In: Medical Imaging 2021: Image Processing, vol. 11596, pp. 597–602. International Society for Optics and Photonics, SPIE (2021)
15. Loshchilov, I., Hutter, F.: Decoupled weight decay regularization. In: International Conference on Learning Representations (2019)
16. Masutani, E.M., Bahrami, N., Hsiao, A.: Deep learning single-frame and multi-frame super-resolution for cardiac MRI. Radiology **295**(3), 552–561 (2020)
17. Pham, C., Ducournau, A., Fablet, R., Rousseau, F.: Brain MRI super-resolution using deep 3D convolutional networks. In: 2017 IEEE 14th International Symposium on Biomedical Imaging (ISBI 2017), pp. 197–200 (2017)
18. Pham, D.L., Xu, C., Prince, J.L.: Current methods in medical image segmentation. Ann. Rev. Biomed. Eng. **2**(1), 315–337 (2000)
19. Smith, L.N.: A disciplined approach to neural network hyper-parameters: part 1-learning rate, batch size, momentum, and weight decay. arXiv preprint arXiv:1803.09820 (2018)
20. Wang, S.L., et al.: Thalamus segmentation using convolutional neural networks. In: Medical Imaging 2021: Image Processing, vol. 11596. International Society for Optics and Photonics, SPIE (2021)
21. Zhang, Y., Li, K., Li, K., Wang, L., Zhong, B., Fu, Y.: Image super-resolution using very deep residual channel attention networks. In: Proceedings of the European Conference on Computer Vision (ECCV), September 2018
22. Zhao, C., Dewey, B.E., Pham, D.L., Calabresi, P.A., Reich, D.S., Prince, J.L.: SMORE: a self-supervised anti-aliasing and super-resolution algorithm for MRI using deep learning. IEEE Trans. Med. Imaging **40**, 805–817 (2020)
23. Zhou, Z., Rahman Siddiquee, M.M., Tajbakhsh, N., Liang, J.: UNet++: a nested U-net architecture for medical image segmentation. In: Stoyanov, D., et al. (eds.) DLMIA/ML-CDS -2018. LNCS, vol. 11045, pp. 3–11. Springer, Cham (2018). https://doi.org/10.1007/978-3-030-00889-5_1

Super-Resolution by Latent Space Exploration: Training with Poorly-Aligned Clinical and Micro CT Image Dataset

Tong Zheng[1(✉)], Hirohisa Oda[2], Yuichiro Hayashi[1], Shota Nakamura[2], Masahiro Oda[1,3], and Kensaku Mori[1,4,5]

[1] Graduate School of Informatics, Nagoya University, Nagoya, Japan
tzheng@mori.m.is.nagoya-u.ac.jp

[2] Nagoya University Graduate School of Medicine, Nagoya, Japan

[3] Information Strategy Office, Information and Communications, Nagoya University, Nagoya, Japan

[4] Information Technology Center, Nagoya University, Nagoya, Japan

[5] Research Center for Medical Big Data, National Institute of Informatics, Tokyo, Japan

Abstract. This paper proposes a super-resolution (SR) method, for performing SR on a poorly-aligned dataset. Super-resolution methods commonly needs aligned low-resolution (LR) and high-resolution (HR) images for training. For obtaining paired LR and HR images in medical imaging, we need to align low and high-resolution data using image registration technology. However, since the hardness of aligning LR and HR images, the aligned LR-HR dataset is always low quality. Conventional SR methods always fail to train using poorly-aligned datasets since these methods need high-quality LR-HR datasets. To tackle this problem, we propose a two-step framework for SR using poorly-aligned datasets. In the first step, we decompose image representation into two parts: one is a content code that captures the image content; the other is a style code that captures the image style and anatomy difference between LR / HR images. To perform SR of a given LR image, we input the content code and a latent variable simultaneously into the SR network to obtain an SR result. In the second step, using the trained SR network and an LR image, we search for a content code, and a style code for generating the most proper SR image. This is conducted by searching for the best content code and the best style code by latent space exploration. We conducted experiments using a poorly-aligned clinical-micro CT lung specimen dataset. Experimental results illustrated the proposed method outperformed conventional SR methods by increasing SSIM from 0.309 to 0.312, and have much more convincing perceptual quality than conventional SR methods.

Keywords: Super-resolution · Poor-quality data processing · Latent space exploration · Lung micro-anatomy

D. Svoboda et al. (Eds.): SASHIMI 2021, LNCS 12965, pp. 24–33, 2021.
https://doi.org/10.1007/978-3-030-87592-3_3

1 Introduction

Super-resolution (SR) is a common image processing topic [16]. The SR aims to reconstruct a high-resolution (HR) image from its corresponding low-resolution (LR) counterpart. The HR images are commonly larger than the LR image and have more anatomical information, such as clearer bronchus and vein in chest CT images. Due to hardware and physics limitations, HR medical imaging comes at the cost of long scan time, and small spatial coverage [2]. Thus, SR of LR images is an optional approach since SR could obtain HR images from LR images.

Previous SR methods [3,4,13] commonly need aligned LR-HR images to train a fully convolutional network (FCN) for SR. Dong *et al.* [3] proposed a deep neural network-based SR method for single-image SR. Lim *et al.* [13] proposed an enhanced deep residual [5] network for SR. Haris *et al.* [4] proposed a network that exploits iterative up- and down-sampling layers for SR. However, a common disadvantage of the above methods is that these methods all need finely aligned LR-HR images for training. In these methods, LR images are acquired by downsampling the HR images using interpolation algorithms such as bicubic [10] interpolation.

On the other hand, in medical imaging, LR and HR images are commonly acquired from various imaging devices. For obtaining paired LR-HR images, image registration is needed. However, precise registration between LR-HR medical images remains challenging [15]. We often obtain only poorly-aligned LR-HR images. Thus, an SR method that could train on poorly-aligned LR-HR images is desired.

We propose an SR network that names cross-modality latent space exploration SR network (CML-SRNet), training only with poorly-aligned LR-HR images. We assume that an image could be decomposed into two parts: one is a content code (Ccode) that captures the content of the image, the other is a style code (Scode) that captures the image style and anatomy difference between LR and HR images. Based on this assumption, we designed an SR framework that only needs poorly-aligned LR-HR images for training. The proposed SR framework contains two steps: the first step trains a CML-SRNet to decompose LR/HR images into Ccode and Scode; the second step is: finds the best Ccode and Scode of the inputted LR images that could perform SR of the LR image by trained CML-SRNet. We constructed a poorly-aligned clinical CT-micro CT images dataset and conducted experiments using this dataset. We evaluated using precisely aligned clinical CT-micro CT image pairs. Note that we use the word “clinical CT images” as CT images taken conventionally at the hospital. The micro CT is another CT modality; the isotropic resolution of micro CT volumes is usually $50 \times 50 \times 50$ μm^3/voxel or higher. Micro CT volumes obtained by a micro CT scanner of resected lung cancer specimens can capture detailed anatomical structures.

The contributions of this paper are: 1) Super-resolution method trained using poorly-aligned LR-HR images; 2) Introducing latent space exploration into SR and outperformed supervised methods and 3) Constructing clinical CT-micro CT dataset of resected lung specimen.

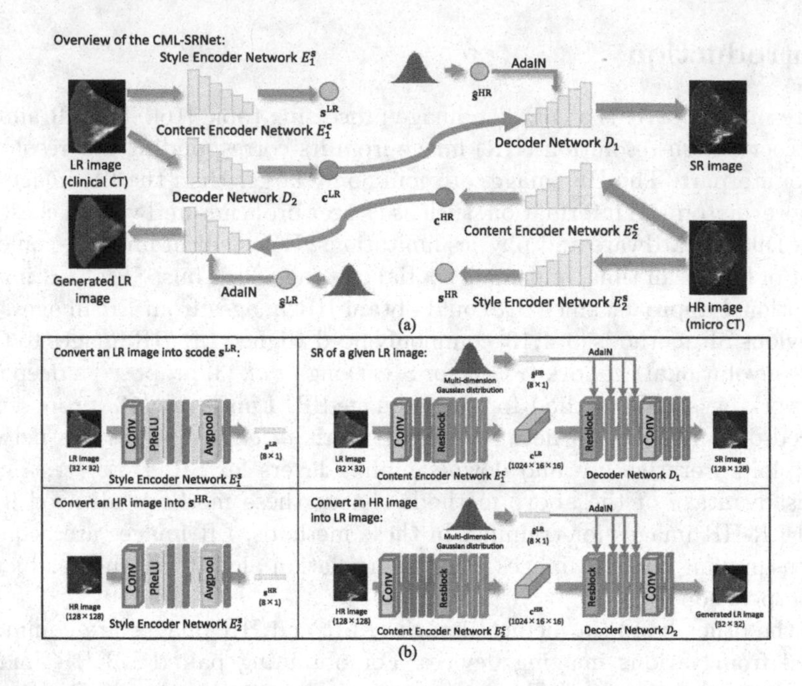

Fig. 1. The structure of CML-SRNet. (a): overview of the CML-SRNet. (b): precise structure of each subnet of CML-SRNet. We utilize poorly-aligned LR-HR images for training. Note that aligned LR-HR images have the same content code (Ccode) but different style code (Scode).

2 Methods

2.1 Overview

Our proposed method performs SR of a given LR image. The proposed method consists of two steps. The first step trains the CML-SRNet using poorly-aligned LR-HR image pairs. Input of the CML-SRNet is an LR image and a latent variable. Output of the CML-SRNet is an SR image. The second step finds an optimal SR result using the trained CML-SRNet. Input of the second step is an LR image, and output of the second step is an optimized SR image.

2.2 CML-SRNet

Given a pair of poorly-aligned LR-HR images, we assume that the LR and HR images could be decomposed into two latent representations: one is a CCode that captures the domain-invariant characteristics of both LR and HR image. The other is an SCode that captures the domain-specific characteristics of LR and HR image respectively. The domain-specific characteristics consist of image deformation, registration error due to poor alignment, etc. Thus, the CCode of paired LR and SR images are the same, while SCode is different. Due to this assumption, we propose the CML-SRNet, to perform super-resolution of LR images while trained on poorly-aligned LR-HR dataset.

The structure of CML-SRNet is shown in Fig. 1. First, a style encoder network $E_1^{\mathbf{s}}$ encodes an LR image $\boldsymbol{x}^{\mathrm{LR}}$ into Scode $\mathbf{s}^{\mathrm{LR}} = E_1^{\mathbf{s}}(\boldsymbol{x}^{\mathrm{LR}})$. Simultaneously, a content code encoder $E_1^{\mathbf{c}}$ encodes the LR image into Ccode $\mathbf{c}^{\mathrm{LR}} = E_1^{\mathbf{c}}(\boldsymbol{x}^{\mathrm{LR}})$. Next, the Ccode $\mathbf{c}^{\mathrm{LR}}$ will be inputted into a decoder network D_1. D_1 is an FCN with residual blocks. The residual blocks are equipped with adaptive instance normalization (AdaIN) [6]. An Scode $\hat{\mathbf{s}}^{\mathrm{HR}}$ sampled from a multidimensional normal distribution $\mathcal{N}_k(0,1)$, passes through a multilayer perceptron (MLP) and generates parameters of the AdaIN layers. Finally, we obtain a HR output $\hat{\mathbf{x}}^{\mathrm{HR}} = D_1(\mathbf{c}^{\mathrm{LR}}, \hat{\mathbf{s}}^{\mathrm{HR}})$. Note that $\hat{\mathbf{x}}^{\mathrm{HR}}$ is actually the SR result $\boldsymbol{x}^{\mathrm{SR}}$ we desire. Sampling a code from a normal distribution is inspired by StyleGAN [8]. Symmetrically, we use a style encoder network $E_2^{\mathbf{s}}$ to encode a HR image $\boldsymbol{x}^{\mathrm{HR}}$ into Scode $\mathbf{s}^{\mathrm{HR}} = E_2^{\mathbf{s}}(\boldsymbol{x}^{\mathrm{HR}})$, and use a content encoder network $E_2^{\mathbf{c}}$ to encode $\boldsymbol{x}^{\mathrm{HR}}$ into Ccode $\mathbf{c}^{\mathrm{HR}} = E_2^{\mathbf{c}}(\boldsymbol{x}^{\mathrm{HR}})$. Then $\mathbf{c}^{\mathrm{HR}}$ and an Scode $\hat{\mathbf{s}}^{\mathrm{LR}}$ sampled from a K-dimensional normal distribution $\mathcal{N}_K(0,1)$ are inputted into a decoder network D_2 to output an LR image $\hat{x}^{\mathrm{LR}}$.

Our loss function contains a *code loss* that ensures the Ccode of paired LR and HR images to be the same and a *SR loss* that ensures the SR image $\boldsymbol{x}^{\mathrm{SR}}$ to be similar enough with its HR counterpart $\boldsymbol{x}^{\mathrm{HR}}$. Other loss terms contains *adversarial loss* and *consistency loss* inspired by MUNIT [7].

The *code loss* is formulated as,

$$\mathcal{L}_{\mathrm{code}} = \mathbb{E}_{\boldsymbol{x}^{\mathrm{LR}} \sim \mathbb{X}^{\mathrm{LR}}, \boldsymbol{x}^{\mathrm{HR}} \sim \mathbb{X}^{\mathrm{HR}}}[\| E_1^{\mathbf{c}}(\boldsymbol{x}^{\mathrm{LR}}) - E_2^{\mathbf{c}}(\boldsymbol{x}^{\mathrm{HR}}) \|_2^2], \tag{1}$$

where $\| \|_2^2$ is the square of 2-norm, $\mathbb{X}^{\mathrm{LR}}$ is the domain of LR images $\boldsymbol{x}^{\mathrm{LR}}$, and $\mathbb{X}^{\mathrm{HR}}$ is the domain of HR images $\boldsymbol{x}^{\mathrm{HR}}$. This loss term is based on an assumption that, while poorly-aligned LR and HR images have differences, they share the same anatomy structure: e.g. a deformed, shriveled blood vessel in LR image while a rounded, plump blood vessel in HR image. The *code loss* prevents arbitrary Ccode from being generated by encoders $E_1^{\mathbf{c}}$ and $E_2^{\mathbf{c}}$.

The *SR loss* is formulated as,

$$\mathcal{L}_{\mathrm{SR}} = \mathbb{E}_{\boldsymbol{x}^{\mathrm{LR}} \sim \mathbb{X}^{\mathrm{LR}}, \boldsymbol{x}^{\mathrm{HR}} \sim \mathbb{X}^{\mathrm{HR}}, \hat{\mathbf{s}}^{\mathrm{HR}} \sim \mathcal{N}_k(0,1)}[\| \boldsymbol{x}^{\mathrm{HR}} - D_1(E_1^{\mathbf{c}}(\boldsymbol{x}^{\mathrm{LR}}), \hat{\mathbf{s}}^{\mathrm{HR}}) \|_2^2], \tag{2}$$

where all the notations are mentioned above. This loss term is commonly used in supervised super-resolution: given an LR image $\boldsymbol{x}^{\mathrm{LR}}$ as input, we desire its SR output $\boldsymbol{x}^{\mathrm{SR}} = D_1(E_1^{\mathbf{c}}(\boldsymbol{x}^{\mathrm{LR}}), \hat{\mathbf{s}}^{\mathrm{HR}})$ to be the same as corresponding HR image $\boldsymbol{x}^{\mathrm{HR}}$. Note that the input of decoder D_1 do not contain $\boldsymbol{x}^{\mathrm{LR}}$'s style code $\mathbf{s}^{\mathrm{LR}}$, because $\boldsymbol{x}^{\mathrm{LR}}$ and $\boldsymbol{x}^{\mathrm{HR}}$ do not share the same Scode domain.

The *adversarial loss* is formulated as,

$$\begin{aligned} \mathcal{L}_{\mathrm{Adv}} = & \mathbb{E}_{\boldsymbol{x}^{\mathrm{LR}} \sim \mathbb{X}^{\mathrm{LR}}, \hat{\mathbf{s}}^{\mathrm{HR}} \sim \mathcal{N}_k(0,1)}[(1 - C_1(D_1(E_1^{\mathbf{c}}(\boldsymbol{x}^{\mathrm{LR}}), \hat{\mathbf{s}}^{\mathrm{HR}})))] \\ & + \mathbb{E}_{\boldsymbol{x}^{\mathrm{HR}} \sim \mathbb{X}^{\mathrm{HR}}, \hat{\mathbf{s}}^{\mathrm{LR}} \sim \mathcal{N}_k(0,1)}[(1 - C_2(D_2(E_2^{\mathbf{c}}(\boldsymbol{x}^{\mathrm{HR}}), \hat{\mathbf{s}}^{\mathrm{LR}})))] \\ & + \mathbb{E}_{\boldsymbol{x}^{\mathrm{HR}} \sim \mathbb{X}^{\mathrm{HR}}}[C_1(\boldsymbol{x}^{\mathrm{HR}})], \\ & + \mathbb{E}_{\boldsymbol{x}^{\mathrm{LR}} \sim \mathbb{X}^{\mathrm{LR}}}[C_1(\boldsymbol{x}^{\mathrm{LR}})], \end{aligned} \tag{3}$$

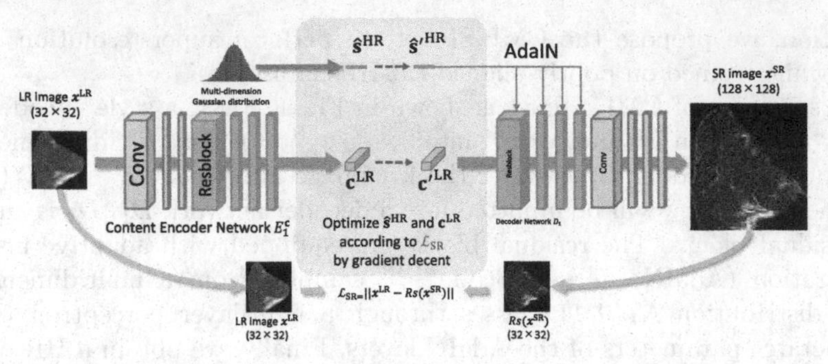

Fig. 2. We explore the space of Ccode and Scode to obtain the final SR result $\boldsymbol{x}^{\mathrm{SR}}$. All notations in this figure are defined in Sect. 2.

where C_1 is a convolutional neural network (CNN)-based classifier that tries to distinguish between SR images $\boldsymbol{x}^{\mathrm{SR}} = D_1(E_1^{\mathbf{c}}(\boldsymbol{x}^{\mathrm{LR}}), \hat{\mathbf{s}}^{\mathrm{HR}})$ and real HR images $\boldsymbol{x}^{\mathrm{HR}}$. We also design a CNN-based classifier C_2 to distinguish LR images $\boldsymbol{x}^{\mathrm{LR}}$ and the fake LR image $\hat{\mathbf{x}}^{\mathrm{LR}}$.

The *consistency loss* is formulated as,

$$\begin{aligned}\mathcal{L}_{\mathrm{Cis}} &= \mathbb{E}_{\boldsymbol{x}^{\mathrm{LR}} \sim \mathbb{X}^{\mathrm{LR}}}[\|D_2(E_1^{\mathbf{c}}(\boldsymbol{x}^{\mathrm{LR}}), E_1^{\mathbf{s}}(\boldsymbol{x}^{\mathrm{LR}})) - \boldsymbol{x}^{\mathrm{LR}}\|_1] \\ &+ \mathbb{E}_{\mathbf{c}^{\mathrm{LR}} \sim p(\mathbf{c}^{\mathrm{LR}}), \hat{\mathbf{s}}^{\mathrm{HR}} \sim p(\hat{\mathbf{s}}^{\mathrm{HR}})}[\|E_2^{\mathbf{c}}(D_1(\mathbf{c}^{\mathrm{LR}}, \hat{\mathbf{s}}^{\mathrm{HR}})) - \mathbf{c}^{\mathrm{LR}}\|_1] \\ &+ \mathbb{E}_{\mathbf{c}^{\mathrm{LR}} \sim p(\mathbf{c}^{\mathrm{LR}}), \hat{\mathbf{s}}^{\mathrm{HR}} \sim p(\hat{\mathbf{s}}^{\mathrm{HR}})}[\|E_2^{s}(D_1(\mathbf{c}^{\mathrm{LR}}, \hat{\mathbf{s}}^{\mathrm{HR}})) - \hat{\mathbf{s}}^{\mathrm{HR}}\|_1],\end{aligned} \tag{4}$$

where $\mathbf{c}^{\mathrm{LR}} = E_1^{\mathbf{c}}(\boldsymbol{x}^{\mathrm{LR}})$ is the Ccode of $\boldsymbol{x}^{\mathrm{LR}}$ and $p(\mathbf{c}^{\mathrm{LR}})$ is the prior of $\mathbf{c}^{\mathrm{LR}}$. $\hat{\mathbf{s}}^{\mathrm{LR}}$ is an Scode sampled from a k-dimensional normal distribution $\mathcal{N}_k(0, 1)$ and $p(\mathbf{c}^{\mathrm{LR}})$ is the prior $\mathcal{N}_k(0, 1)$. The consistency loss is designed to ensure encoder E_1 and decoder D_1 are inverse to each other, as well as E_2 and decoder D_2. Given the Ccode and Scode of $\boldsymbol{x}^{\mathrm{LR}}$, the *consistency loss* ensures $\boldsymbol{x}^{\mathrm{LR}}$ could be reconstructed by the Ccode and Scode. Moreover, the *consistency loss* also ensures the Ccode $\mathbf{c}^{\mathrm{LR}}$ and Scode $\mathbf{s}^{\mathrm{LR}}$ could be reconstructed. This helps to keep the cycle consistency [7] of the CML-SRNet and makes training more stable.

The total loss function is formulated as,

$$\mathcal{L}_{\mathrm{Total}} = \lambda_{\mathrm{code}}\mathcal{L}_{\mathrm{code}} + \lambda_{\mathrm{SR}}\mathcal{L}_{\mathrm{SR}} + \lambda_{\mathrm{Adv}}\mathcal{L}_{\mathrm{Adv}} + \lambda_{\mathrm{Cis}}\mathcal{L}_{\mathrm{Cis}}, \tag{5}$$

where λ_{code}, λ_{SR}, λ_{Adv} and λ_{Cis} are weights of each loss term.

2.3 SR by Latent Space Exploration

Given an LR image $\boldsymbol{x}^{\mathrm{LR}}$, our goal is to find its HR counterpart $\boldsymbol{x}^{\mathrm{HR}}$. This problem could be formulated as minimizing such function,

$$\mathcal{L}_{\mathrm{SR}} = \mathbb{E}_{\boldsymbol{x}^{\mathrm{HR}} \sim \mathbb{X}^{\mathrm{HR}}}[\|\boldsymbol{x}^{\mathrm{HR}} - \boldsymbol{x}^{\mathrm{SR}}\|_2^2], \tag{6}$$

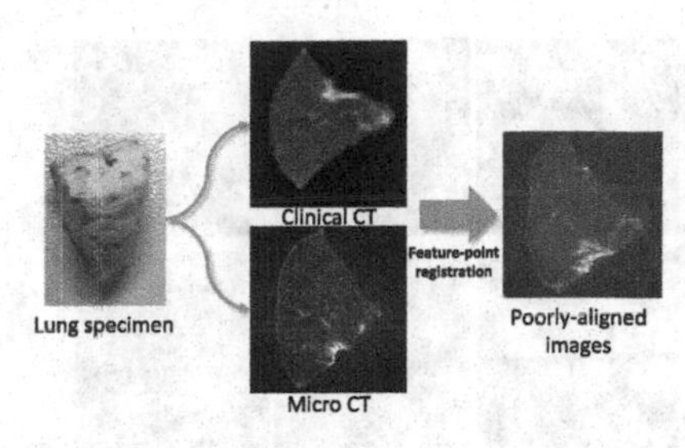

Fig. 3. Preparation of dataset. We scanned lung specimens by clinical CT and micro CT respectively. Then we mark feature points (red points) manually to align clinical CT to micro CT.

Table 1. Size and resolution of each case.

Lung number	Clinical CT size / resolution	Micro CT size / resolution
Lung-A	65 × 65 × 35 voxels / 0.549 × 0.549 × 0.5 mm³	1024 × 1024 × 545 voxels / 39 × 39 × 39 μm³
Lung-B	80 × 80 × 110 voxels / 0.549 × 0.549 × 0.5 mm³	1024 × 1024 × 1078 voxels / 52 × 52 × 52 μm³
Lung-C	80 × 100 × 66 voxels / 0.549 × 0.549 × 0.5 mm³	1024 × 1024 × 1093 voxels / 52 × 52 × 52 μm³
Lung-D	90 × 90 × 100 voxels / 0.549 × 0.549 × 0.5 mm³	1024 × 1024 × 1087 voxels / 52 × 52 × 52 μm³
Lung-E	80 × 80 × 80 voxels / 0.549 × 0.549 × 0.5 mm³	1024 × 1024 × 1086 voxels / 52 × 52 × 52 μm³

and since $\boldsymbol{x}^{\mathrm{SR}} = D_1(E_1^{\mathbf{c}}(\boldsymbol{x}^{\mathrm{LR}}), \hat{\mathbf{s}}^{\mathrm{HR}}) = D_1(\mathbf{c}^{\mathrm{LR}}, \hat{\mathbf{s}}^{\mathrm{HR}})$ as illustrated in Sect. 2.2, the above function could be rewrote as,

$$\mathcal{L}_{\mathrm{SR}} = \mathbb{E}_{\boldsymbol{x}^{\mathrm{HR}} \sim \mathbb{X}^{\mathrm{HR}}, \mathbf{c}^{\mathrm{LR}} \sim p(\mathbf{c}^{\mathrm{LR}}), \hat{\mathbf{s}}^{\mathrm{HR}} \sim p(\hat{\mathbf{s}}^{\mathrm{HR}})}[\|\boldsymbol{x}^{\mathrm{HR}} - D_1(\mathbf{c}^{\mathrm{LR}}, \hat{\mathbf{s}}^{\mathrm{HR}})\|_2^2], \tag{7}$$

note that decoder D_1 is trained and cannot be optimized anymore. Thus, we could only optimize the Ccode $\mathbf{c}^{\mathrm{LR}}$ and Scode $\hat{\mathbf{s}}^{\mathrm{HR}}$ to minimize the above function. The problem is, a critical variable $\boldsymbol{x}^{\mathrm{HR}}$ remains unknown. And since

$$\|\boldsymbol{x}^{\mathrm{HR}} - D_1(\mathbf{c}^{\mathrm{LR}}, \hat{\mathbf{s}}^{\mathrm{HR}})\|_2^2 \approx \|Rs(\boldsymbol{x}^{\mathrm{HR}}) - Rs(D_1(\mathbf{c}^{\mathrm{LR}}, \hat{\mathbf{s}}^{\mathrm{HR}}))\|_2^2, \tag{8}$$

where Rs is a resampling function. If Rs downsamples $\boldsymbol{x}^{\mathrm{HR}}$ to the same size as $\boldsymbol{x}^{\mathrm{LR}}$, then $Rs(\boldsymbol{x}^{\mathrm{HR}}) \approx \boldsymbol{x}^{\mathrm{LR}}$. Thus, as an alternative, we minimize such function,

$$\mathcal{L}_{\mathrm{SR}} = \mathbb{E}_{\boldsymbol{x}^{\mathrm{HR}} \sim \mathbb{X}^{\mathrm{HR}}, \mathbf{c}^{\mathrm{LR}} \sim p(\mathbf{c}^{\mathrm{LR}}), \hat{\mathbf{s}}^{\mathrm{HR}} \sim p(\hat{\mathbf{s}}^{\mathrm{HR}})}[\|\boldsymbol{x}^{\mathrm{LR}} - Rs(D_1(\mathbf{c}^{\mathrm{LR}}, \hat{\mathbf{s}}^{\mathrm{HR}}))\|_2^2], \tag{9}$$

and note that the size of $Rs(D_1(\mathbf{c}^{\mathrm{LR}}, \hat{\mathbf{s}}^{\mathrm{HR}}))$ is the same as $\boldsymbol{x}^{\mathrm{LR}}$. We use stochastic gradient descent (SGD) to optimize $\mathbf{c}^{\mathrm{LR}}$ and $\hat{\mathbf{s}}^{\mathrm{HR}}$ as to minimize the above function. After optimization, we obtain optimized $\mathbf{c}'^{\mathrm{LR}}$ and $\hat{\mathbf{s}}'^{\mathrm{HR}}$. The final SR output is $\boldsymbol{x}^{\mathrm{SR}} = D_1(\mathbf{c}'^{\mathrm{LR}}, \hat{\mathbf{s}}'^{\mathrm{HR}})$.

3 Experiments and Results

3.1 Dataset and Parameter Settings

Figure 3 illustrates the preparation of the dataset. Table 1 illustrates the size/resolution of the clinical CT and micro CT images in the dataset. Since there is no existing poorly-aligned LR-HR dataset, we prepared a poorly-aligned clinical CT (HR) - micro CT (LR) dataset ourselves. We prepared five resected lung specimens resected from lung cancer patients, and used the Heitzman method [14] to fix the lung specimens. We took clinical CT volumes of these lung specimens using a clinical CT scanner (SOMATOM Definition Flash, Siemens Inc.,

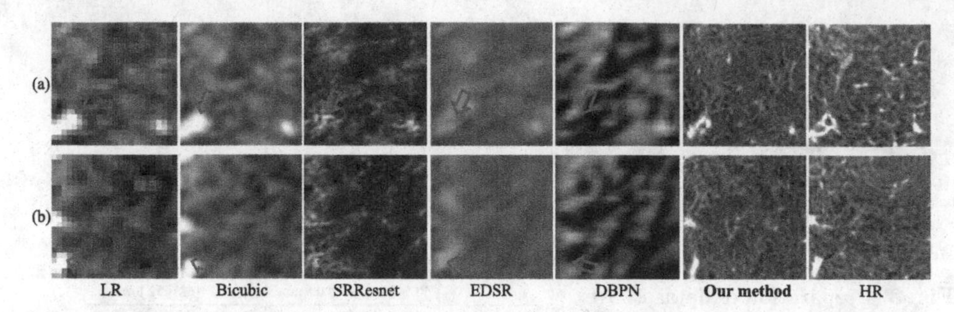

Fig. 4. Qualitative results (8×SR) of proposed CML-SRNet and previous results [4, 12, 13]. Since we trained each method on a poorly-aligned LR-HR dataset, all previous methods performed badly except ours. Row (a): images of bronchioles. Row (b): images of arteries. Note LR images are cropped from the specimen's clinical CT and HR image cropped from micro CT. Recommend zooming in to watch places marked by red arrows. (Color figure online)

Munich) in the resolution of $549 \times 549 \times 500\ \mu\text{m}^3$/voxel. We took micro CT volumes of these specimens using a micro CT scanner (inspeXio SMX90CT Plus, Shimadzu, Kyoto) in the resolution of $(42-52) \times (42-52) \times (42-52)\ \mu\text{m}^3$/voxel. For aligning micro CT and clinical CT images, We chose 6–10 feature points of each case manually and used affine transformation [1] to obtain a poorly-aligned clinical (LR) - micro CT (HR) dataset. We also used non-rigid registration [9] to precisely align one clinical-micro CT case for quantitative evaluation.

For training the CML-SRNet, we randomly cropped 2000 2D patches from each case. Since we have 4 cases for training, the total number of patches for training is 8000. We experimented 4× and 8× SR. Considering memory size and size of CT volume, the size of patches extracted from the clinical CT volumes was 32×32 pixels for 4× SR and 24×24 pixels for 8× SR. The size of patches extracted from the micro CT volumes was 128×128 pixels for 4× SR and 192×192 pixels for 8× SR. We set the network learning rate to 10^{-4}, and halves every 30 epochs. The epoch number is 200, and optimize algorithm is Adam [11]. The size of Scode is 8 (sampled from 8-dimensional normal distribution). Weight of each loss term: $\lambda_{\text{code}} = 0.2$, $\lambda_{\text{SR}} = 1.0$, $\lambda_{\text{Adv}} = 1.0$, $\lambda_{\text{Cis}} = 1.0$. About finding $\mathbf{c}'^{\text{LR}}$ and $\hat{\mathbf{s}}'^{\text{HR}}$ in latent space exploration, we set the iteration number to 100 and set SGD as the optimization algorithm with learning rate= 0.1 as to obtain $\mathbf{c}'^{\text{LR}}$ and $\hat{\mathbf{s}}'^{\text{HR}}$. All algorithms and networks were implemented by Python 3.8.6 and PyTorch 1.7.0 on Nvidia Quadro P6000. The parameter number of CML-SRNet is 391,032,342.

3.2 Results

Figure 4 and 5 show the result. We compared our method with supervised SR methods such as SRResnet [12], EDSR [13] and DBPN [4]. Note that SRResnet, EDSR and DBPN are all trained on the poorly-aligned clinical CT - micro CT dataset. We also compared our method with bicubic interpolation in Fig. 4.

Table 2. Performance of proposed CML-SRNet, and previous methods [4,12,13]

Methods	PSNR (dB)	SSIM
Bicubic	11.143	0.50
SRResnet [12]	13.767	0.309
EDSR [13]	16.886	0.233
DBPN [4]	14.306	0.290
Our method	16.117	0.312

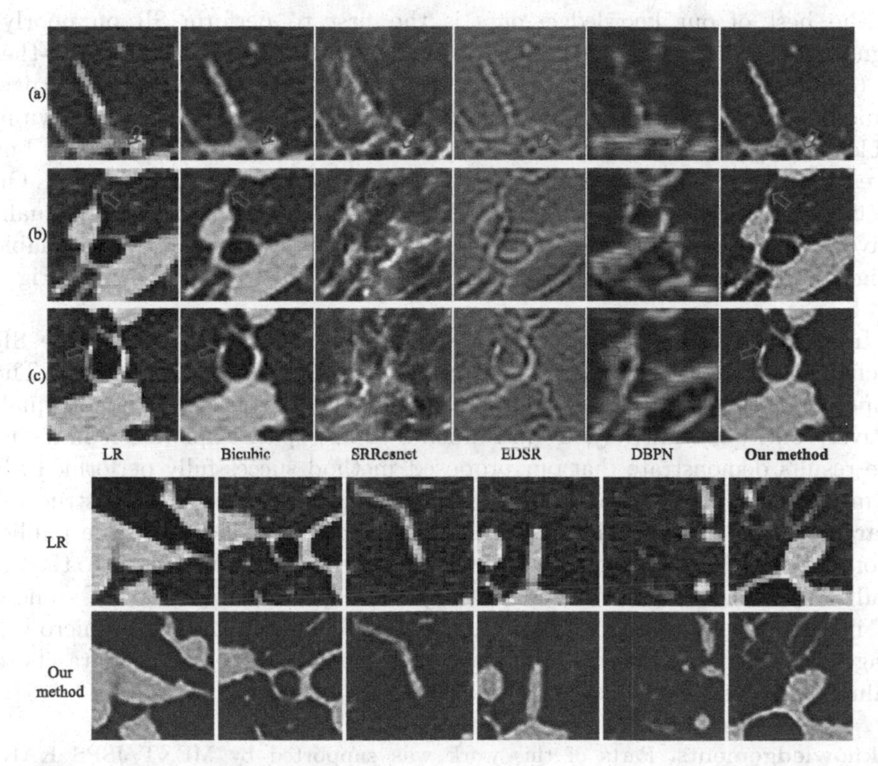

Fig. 5. Qualitative results of real clinical CT (4×SR). Above: Row (a) edge of tumor, row (b) bronchus and row (c) more bronchus. Supervised methods which needs finely aligned LR-HR image pairs for training (SRResnet, EDSR, DBPN) do not work at all on our poorly-aligned dataset. On the other hand, our CML-SRNet has the best Qualitative result. Recommend zooming in to watch places marked by red arrows. Below: more results of our proposed method.

For quantitative evaluation, we evaluated our method using 500 precisely-aligned clinical-micro CT image pairs. The quantitative results are shown in Table 2. For quantitative evaluation, we evaluated our method using 500 clinical CT images from a lung cancer patient. Our method outputted satisfying SR result while supervised methods such as SRResnet and EDSR outputted very blurry and nearly arbitrary results. Our method also reconstructed microanatomical structures such as small bronchus and edge of tumor better than previous methods, as in Fig. 5. We utilize PSNR and SSIM because other metrics such as LPIPS [17] are designed for photographic images and not sufficient for our dataset. Proposed CML-SRNet also outperformed these previous methods quantitatively (by SSIM).

4 Discussion and Conclusion

To the best of our knowledge, ours is the first to perform SR on poorly-aligned LR-HR dataset. Our proposed CML-SRNet outperformed state-of-the-art (SOTA) supervised SR methods [4,12,13]. Note that CML-SRNet outperformed all these methods in SSIM but has lower PSNR than EDSR, as shown in Table 2. We consider this because SR images outputted by EDSR are blurry but have closer average intensity with the HR image, resulting in a higher PSNR. On the other hand, the proposed CML-SRNet outperformed these methods qualitatively. SOTA supervised SR methods outputted very blurry or unreasonable images while proposed CML-SRNet correctly performed SR as shown in Fig. 4 and 5.

In this paper, we proposed a novel SR approach for performing the SR of clinical CT images, while training on poorly-aligned LR-HR dataset. The proposed method CML-SRNet outperformed SOTA supervised method qualitatively and quantitatively, by introducing latent space exploration into SR. The results demonstrate that our proposed method successfully performed SR of lung clinical CT eight-times by height and width. We consider constructing a larger dataset by collecting more lung specimens and open it to the public. Another issue is our method introduces some micro CT-like artifacts into the SR result. We consider this is because of the poor alignment of clinical CT - micro CT images; thus aligning the images more precisely and denoising of micro CT images is needed. We also plan to verify our method on other public datasets to evaluate its consistency.

Acknowledgements. Parts of this work was supported by MEXT/JSPS KAKENHI (26108006, 17H00867, 17K20099), the JSPS Bilateral International Collaboration Grants, the AMED (JP19lk1010036 and JP20lk1010036) and the Hori Sciences & Arts Foundation.

References

1. Bebis, G., Georgiopoulos, M., da Vitoria Lobo, N., Shah, M.: Learning affine transformations. Pattern Recogn. **32**(10), 1783–1799 (1999). https://doi.org/10.1016/S0031-3203(98)00178-2
2. Chen, Y., Shi, F., Christodoulou, A.G., Xie, Y., Zhou, Z., Li, D.: Efficient and accurate MRI super-resolution using a generative adversarial network and 3D multi-level densely connected network. In: Frangi, A.F., Schnabel, J.A., Davatzikos, C.,

Alberola-López, C., Fichtinger, G. (eds.) MICCAI 2018. LNCS, vol. 11070, pp. 91–99. Springer, Cham (2018). https://doi.org/10.1007/978-3-030-00928-1_11
3. Dong, C., Loy, C.C., He, K., Tang, X.: Image super-resolution using deep convolutional networks. IEEE Trans. Pattern Anal. Mach. Intell. **38**(2), 295–307 (2015). https://doi.org/10.1109/TPAMI.2015.2439281
4. Haris, M., Shakhnarovich, G., Ukita, N.: Deep back-projection networks for super-resolution. In: Proceedings of the IEEE Conference on Computer Vision and Pattern Recognition, pp. 1664–1673. IEEE (2018). https://doi.org/10.1109/CVPR.2018.00179
5. He, K., Zhang, X., Ren, S., Sun, J.: Deep residual learning for image recognition. In: Proceedings of the IEEE Conference on Computer Vision and Pattern Recognition, pp. 770–778. IEEE (2016). https://doi.org/10.1109/CVPR.2016.90
6. Huang, X., Belongie, S.: Arbitrary style transfer in real-time with adaptive instance normalization. In: Proceedings of the IEEE International Conference on Computer Vision, pp. 1501–1510. IEEE (2017). https://doi.org/10.1109/ICCV.2017.167
7. Huang, X., Liu, M.-Y., Belongie, S., Kautz, J.: Multimodal unsupervised image-to-image translation. In: Ferrari, V., Hebert, M., Sminchisescu, C., Weiss, Y. (eds.) ECCV 2018. LNCS, vol. 11207, pp. 179–196. Springer, Cham (2018). https://doi.org/10.1007/978-3-030-01219-9_11
8. Karras, T., Laine, S., Aila, T.: A style-based generator architecture for generative adversarial networks. In: Proceedings of the IEEE Conference on Computer Vision and Pattern Recognition, pp. 4401–4410. IEEE (2019). https://doi.org/10.1109/CVPR.2019.00453
9. Keszei, A.P., Berkels, B., Deserno, T.M.: Survey of Non-Rigid Registration Tools in Medicine. J. Digit. Imaging **30**(1), 102–116 (2016). https://doi.org/10.1007/s10278-016-9915-8
10. Keys, R.: Cubic convolution interpolation for digital image processing. IEEE Trans. Acoust. Speech Signal Process. **29**(6), 1153–1160 (1981). https://doi.org/10.1109/TASSP.1981.1163711
11. Kingma, D.P., Ba, J.: Adam: a method for stochastic optimization. In: Proceedings of the 3rd International Conference on Learning Representations (2014)
12. Ledig, C., et al.: Photo-realistic single image super-resolution using a generative adversarial network. In: Proceedings of the IEEE Conference on Computer Vision and Pattern Recognition, pp. 4681–4690. IEEE (2017). https://doi.org/10.1109/CVPR.2017.19
13. Lim, B., Son, S., Kim, H., Nah, S., Mu Lee, K.: Enhanced deep residual networks for single image super-resolution. In: Proceedings of the IEEE conference on Computer Vision and Pattern Recognition Workshops, pp. 136–144. IEEE (2017). https://doi.org/10.1109/CVPRW.2017.151
14. Markarian, B.: Preparation of inflated lung specimens. The lung: Radiologic-pathologic Correlation, pp. 4–12 (1984). https://doi.org/10.1111/j.1740-8261.1983.tb01539.x
15. Sotiras, A., Davatzikos, C., Paragios, N.: Deformable medical image registration: a survey. IEEE Trans. Med. Imaging **32**(7), 1153–1190 (2013). https://doi.org/10.1109/TMI.2013.2265603
16. Wang, Z., Chen, J., Hoi, S.C.: Deep learning for image super-resolution: a survey. IEEE Trans. Pattern Anal. Mach. Intell. (2020). https://doi.org/10.1109/TPAMI.2020.2982166
17. Zhang, R., Isola, P., Efros, A.A., Shechtman, E., Wang, O.: The unreasonable effectiveness of deep features as a perceptual metric. In: Proceedings of the IEEE Conference on Computer Vision and Pattern Recognition, pp. 586–595 (2018)

A Glimpse into the Future: Disease Progression Simulation for Breast Cancer in Mammograms

Ibrahim Jubran[1,2](✉), Moshiko Raboh[1], Shaked Perek[1], David Gruen[1], and Efrat Hexter[1]

[1] IBM Research, Haifa University, Mount Carmel 31905, Israel
{moshiko.raboh,david.gruen}@ibm.com, {shaked.perek,efrathex}@il.ibm.com
[2] Robotics & Big Data Lab, Computer Science Department, University of Haifa, Haifa, Israel

Abstract. Simulating the disease progression of a suspicious finding in mammography (MG) images could assist in the early detection of breast cancer, where the telltale signs of malignancy are subtle and hard to detect. It could also decrease both unnecessary biopsies and treatments of indolent or low-grade disease that might otherwise remain asymptomatic. We propose a novel approach to simulating disease progression, as a significant step towards those goals. Our architecture uses the powerful Wasserstein GAN in combination with a novel component that simulates the progression of the disease in deep feature space. This allows us to learn from unlabeled longitudinal MG pairs of current and prior studies, stabilize the learning procedure, and overcome misalignment between the MG pairs. Our output image replicates an actual MG image, maintains the prior's shape and general appearance while also containing a finding with characteristics that resemble the current image's suspicious finding. We simulate a progression of: (i) a full MG prior image in low-resolution, and (ii) a high-resolution patch in suspicious areas of the prior image. We demonstrate the effectiveness of our pipeline in achieving the above goals using quantitative and qualitative metrics and a reader study. Our results show the high quality of our simulation and the promise it holds for early risk stratification.

Keywords: Disease progression · Mammography · Generative models

1 Introduction

Breast cancer screening MG is aimed at detecting early signs of malignant lesions. It has been shown in numerous studies that early detection can reduce mortality by roughly 40%, as well as often allowing for less extensive treatments [13]. During a typical screening examination, each breast is compressed between two

M. Raboh, S. Perek—Authors contributed equally.

D. Svoboda et al. (Eds.): SASHIMI 2021, LNCS 12965, pp. 34–43, 2021.
https://doi.org/10.1007/978-3-030-87592-3_4

plates, and images of each breast are acquired in two standardized projections: Craniocaudal (CC) and Mediolateral-Oblique (MLO) views.

Modeling and synthesizing a disease progression (DP) can serve as a powerful tool in the radiology reading rooms. Emphasizing the early signs of breast cancer, which might be subtle and below the threshold of detection, could both improve early discovery and prevent unnecessary biopsies. This can be done by distinguishing benign from malignant processes and by predicting the course of a disease without the need for histopathologic evaluation. Correctly predicting the course of a disease could also minimize the potential harm caused by overdiagnosis and over-treating what would otherwise remain an indolent or asymptomatic condition [13]. Instead, the physician might administer less extensive treatment or optimize the frequency and/or type of screening examinations.

Although precise practices vary in different geographies, women globally undergo MG screening, whether or not they are in a high-risk group. For each patient, this practice creates a longitudinal dataset of prior (past) image and current (newer) image series, acquired from the same patient at different times. These image series have the potential to display the evolution of breast cancer when viewed over time, and can be used to train a model to simulate DP. However, this is a very challenging task. It requires: (i) detecting early signs of disease in an image, (ii) reliably simulating a disease that resembles the one detected in a newer image of the same patient, and (iii) maintaining the original image characteristics. The learning procedure must also be able to overcome any misalignment of the longitudinal image series caused by the non-linear and inconsistent deformations a breast undergoes during MG screening. These deformations remain after an affine registration attempt; see Fig. 1.

Following a disease as it progresses through time is a relatively new and not thoroughly explored concept. One approach is to use Generative Adversarial Networks (GANs) [7] to generate images illustrating the progression of a disease. GANs aim to model a dataset's distribution and are composed of two competing components: a generator that generates a fake image, and a discriminator that distinguishes between real and fake images. It is commonly presumed that GANs can produce higher quality results but are harder to train, as compared to the alternative Variational Auto Encoders (VAEs). Simulating DP has been researched in brain MRI neurodegenerative disease [15,21,22]. Similar to our approach, Wegmayr et al. [21] and Zhao et al. [22], also use GANs. However, to the best of our knowledge, our work is the first to simulate tumor progression in MG images, overcoming challenges of early signs of cancer detection and considerable misalignment in longitudinal MGs. The few prior works that handle MG images focus on either synthesizing random MGs [10] or inserting a virtual lesion into an MG given a tumor mask or an initial seed [11,18].

The following question naturally arises. Can we model the relation between misaligned MG pairs and synthesize an image that simultaneously maintains the attributes of the prior image, while combining them with the properties of a lesion from the current image? The main contributions of this work are:

(i) We affirmatively answer the question above and show that simulating DP in MG is indeed feasible. We demonstrate the desired properties in our synthetic images using qualitative and quantitative metrics, alongside confirmation via a reader study by a radiologist. To our knowledge, this is the first non-trivial attempt for DP on longitudinal MG data.
(ii) We propose a novel architecture based on WGANs that allows to learn from longitudinal pairs in a semi-supervised manner, stabilize the learning procedure, and overcome misalignments between the pairs of images.
(iii) Finally, we show that the same pipeline, with minor modifications, can be used to simulate a high-resolution progression of a lesion in a manually selected patch, with no need for a tumor mask or initial seed as input.

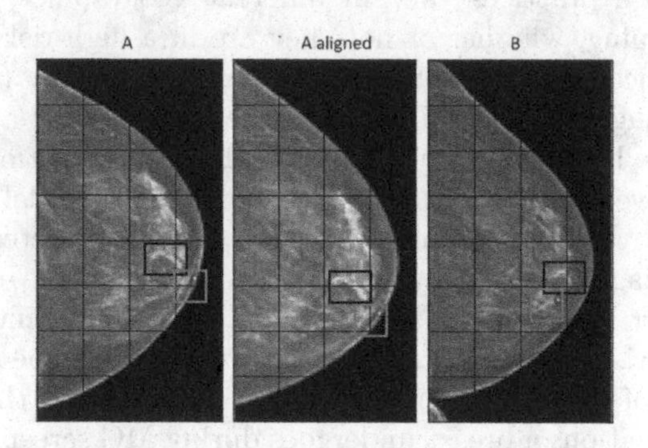

Fig. 1. Illustration of the misalignment challenge introduced by non-linear deformations a breast undergoes during an MG procedure. *A* and *B* are two MG images of the same patient acquired *on the same day*. *A aligned* is the result of an affine registration between images *A* and *B*. The red and blue boxes mark the same observation in all three images. The dataset used in this work consist of MG pairs taken more than 1 day apart, thus the misalignment is even more severe.

2 Methods

This section provides details for the architecture of our proposed model; see Fig. 2. Given a patient's screening history, we can create a cancer development timeline. Our input is thus a dataset of pairs, comprising prior (past) and current (newer) MGs of the same patient. Our goal is to simulate a "progressed prior" (PP) image, based on the prior image. This image simulates what a lesion can develop into, even if it is unnoticed in the prior image. Our output PP should simultaneously: (i) appear as a valid MG image, (ii) contain a finding with properties that resemble the characteristics of the current image's suspicious finding, and (iii) maintain the prior's shape, structure, and general appearance.

2.1 Model Architecture

GANs are a natural selection when modeling and sampling from unknown distributions. Our proposed pipeline uses the state-of-the-art WGAN [3] with Gradient Penalty [8], due to its more stable and robust training process. During training, the generator's input consists of longitudinal image pairs from the same patient. Our pipeline is depicted in Fig. 2. First, the prior image is plugged into the encoder E, to produce a "prior deep feature (DF)" representation. This representation then undergoes a transformation using an additional model M which we call *midnet*, to achieve a progressed DF, which serves as the inputs to the decoder D. The decoder then outputs a PP image; Second, a current image is embedded using the encoder E to produce "current deep features". Both the PP and the current images are each concatenated with the original prior, and are fed to our conditional critic (discriminator) C, as in standard conditional GANs [12]. The conditional WGAN loss in our case is:

$$L_{WGAN} = \mathbb{E}_{(x_p,x_c)\sim p_{data}}(C(x_p, x_c)) - \mathbb{E}_{(x_p,x_c)\sim p_{data}}(C(x_p, G(x_p)) + \lambda \cdot GP \quad (1)$$

where p_{data} is the real paired-data distribution, GP is the gradient penalty term [8], and $\lambda = 10$ is a weighting factor.

In a standard GAN, we would like our generator to output a PP image. This output is conditioned on our prior image, but should resemble its current counterpart image. Hence, a similarity loss between input pairs can be helpful. However, despite initial affine transformation alignment, misalignment between prior and current images still exists due to the non-linear deformations of the breast as in Fig. 1, which make such a simple pixel-wise loss infeasible.

Similarity loss between the DF maps of intermediate layers is more tolerant to misalignment. This occurs because the model can learn a higher-level representation of the image that is agnostic to minor misalignment. We use this DF similarity loss approach by adding the midnet, in between the encoder and decoder models. As input, our midnet receives the prior image's deep features, which are generated by the encoder. It then works to progress those features so they resemble the current image's deep features. Similar approaches have been used in the past [5,20]. We define the deep feature loss as: $L_{DF} = \frac{1}{N}\sum_i \|M(f_{p_i}) - f_{c_i}\|^2$. where $M(f_{p_i})$ and f_{c_i} are the progressed prior DFs and current DFs at the ith level of our model, respectively. Aside from the progression task, the above DF loss serves as a deep-supervision for training M, and for stabilizing the GAN training by enforcing a gradient progression in a desired direction.

For our model to go beyond generating images that resemble the prior image, and better characterize the disease in the image, we added a segmentation model. This model aims to localize the disease in the synthetic image. By combining this model's output along with a ground truth (GT) segmentation map of the current images, we can penalize the progression of images that lack disease characteristics similar to the ones in the current. This step helps the model to better learn the right location for the progressed disease. A segmentation dice loss L_{dice} is used between the segmentation maps, as utilized in [17,19]. The complete model loss is defined by the following weighted sum of loses $L = \alpha_{WGAN} \cdot L_{WGAN} + \alpha_{DF} \cdot$

$L_{DF}+\alpha_{dice}\cdot L_{dice}$. Both L_{DF} and L_{dice} are auxiliary terms that helps our model focus on the disease properties and the progression simulation.

2.2 Semi-supervised Training

For our work, an ideal dataset would be longitudinal data of prior/current pairs with the local suspicious region annotated in the current image. However, such simultaneously paired and annotated data is very costly and rare. In our approach, we pre-train a standard segmentation model similar to [1,17]. The data comprises locally annotated, but *not necessarily longitudinal*, single images. Such data is much more available. Now, the desired segmentation map can be extracted from the current images of the longitudinal data, without the need for longitudinal data with GT annotation.

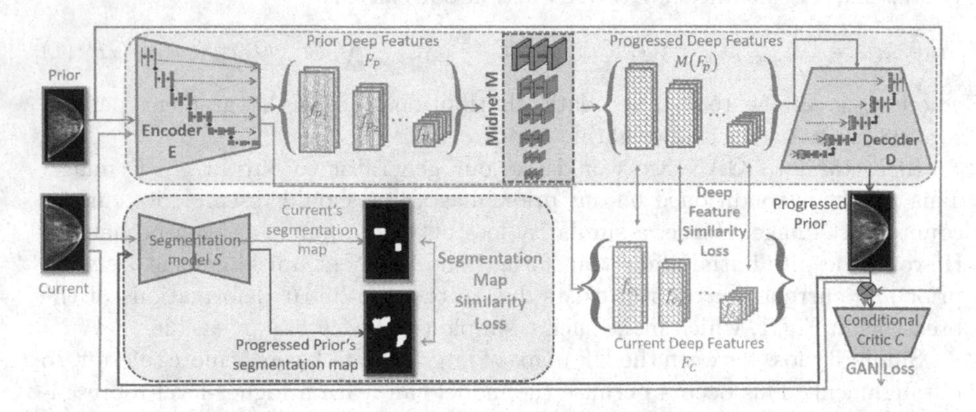

Fig. 2. Training pipeline. A corresponding pair of input images, Prior (I_P) and Current (I_C), are fed into a standard UNet [16] encoder (E) with 5 skip connections [6] to obtain the sets F_P and F_C of deep feature maps. Each set contains the feature maps from all 6 levels of the encoder. Then, each of those feature are plugged into a separate 3 layer CNN to obtain the progressed features $M(F_P)$. The union of those CNNs form our deep feature progression model (Midnet M). Then, $M(F_P)$ is decoded back into a "Progressed Prior" image I_{PP} using a standard UNet decoder (D). The models E, M, and D form our generator, shown in the red box. I_P is then concatenated to both I_C and I_{PP} before they are plugged into the conditional critic C. The critic C resembles the encoder E but followed by a fully connected layer. Both I_C and I_{PP} are also plugged into a segmentation model S to obtain two segmentation maps S_C and S_{PP}; see Sect. 2.2. We applied three losses (illustrated in orange). See details in Sect. 2.1.

3 Experiments

We now give details about our datasets, tasks, training procedure, and results.

Data. We used three private cohorts, each for a different task as explained below: **Longitudinal cohort (i)** consists of 2740 patients (1269 malignant, 1497 benign, 26 have both) from 5 providers with longitudinal MG images of the CC view. **Longitudinal cohort (ii)** consists of 1140 patients (690 malignant, 496 benign, 6 have both) from 5 providers with longitudinal MG images of the CC and MLO views. **A segmentation cohort** used to train the segmentation model; see Sect. 2.2. It consists of 5930 patients (3251 malignant, 2776 benign, 97 have both) from 8 providers, with MG images of CC and MLO views, containing local annotations to mark the contour of malignant and benign tumors. This is not a longitudinal dataset and contains no current images from the previous cohorts. In the longitudinal cohorts, the current image is defined as the most recent image that has a suspicious lesion that underwent a biopsy with a reported outcome. A prior image is an image acquired between 10–36 months before the current image. The cohorts were randomly split into 80% train set and 20% test set. See image pair example in Fig. 1.

Training Details. Our data first undergoes a breast image extraction procedure using an algorithm that zeros out the background and crops the image around the breast. The image values are scaled to $[-1, 1]$. For every generator iteration we trained the discriminator two iterations. We used batch size of 16 and Adam optimizer with $lr = 1e^{-4}$. Our generator uses 3×3 convolution kernels, followed by BN layers and ReLU, except for the output layer of D, which applies Tanh. The critic uses 5×5 convolution kernels, with no BN as recommended in [8]. We add zero-mean Gaussian noise with a standard deviation of 1 to its input [2]. The deviation is reduced by 0.95 every epoch. Downsampling and upsampling layers are strided and transpose convolutions, respectively [14]. For training stability, we applied a similarity loss L_{sim} between the PP image and the current image. The loss coefficients were set to $\alpha_{WGAN} = 100$, $\alpha_{DF} = 5$, $\alpha_{sim} = 5$, and $\alpha_{dice} = 0$ for the first 700 epochs, and $\alpha_{dice} = 10$ for the rest. The training was stopped after 1000 total epochs, and the last model was chosen. We decreased α_{DF} and α_{sim} by x0.95 and x0.5, respectively, every epoch. We used Python 3.6 with PyTorch 1.1, FuseMedML framework [9], and a single NVIDIA Tesla v100 16GB GPU.

Tasks. We tested the two following tasks: **(i) Progression of full image.** Our goal here was to progress a full MG prior image from the longitudinal cohort (i). We aligned every input image to a common template image via affine registration and down-sampled it to 256×128. We maintained the original aspect ratio by padding the larger axis with zeros, as needed. The down-sampling was necessary due to memory constraints. **(ii) Progression of patches.** Breast cancer screening is often based on fine details that can be identified only at the original resolution. We provide the option to simulate high-resolution DP on a selected suspicious patch. Since most patches from either CC/MLO view appear very similar, we used the longitudinal cohort (ii). We aligned every prior image to its corresponding current image via affine registration. Next, we extracted a patch of size 256×256 centered around a lesion annotated in the current image and extracted its corresponding patch in the prior image. To encourage generation of patches that can successfully blend back to the prior image, we take an inner-patch of size 192×192 from the 256×256 patch at the output of the generator, and plug this inner-patch back into the original 256×256 patch before feeding it into the discriminator. As a result, the discriminator can also witness the blending of the PP patch back to the prior image. The extracted patches are small and tumor-centered, hence, we disable the segmentation loss.

Table 1. Evaluation measurements. Upper table: radiologist's evaluation on 20 prior, 10 current and 10 PP images. Lower table: the final quantitative metrics on the test set (estimated using bootstrapping with 10,000 repetitions and 95% confidence interval). All metrics compare the (registered) prior or PP, to the corresponding current image.

	Prior	Current	Progressed Prior
Classified as real	100%	100%	80%
Sent for biopsy	60%	100%	80%

	Full Image		Patches	
	Prior	Progressed Prior	Prior	Progressed Prior
Dice	0.27[0.25-0.28]	**0.37[0.35-0.38]**	0.17[0.15-0.18]	**0.32[0.29-0.34]**
MSE	0.058[0.056-0.061]	**0.03[0.028-0.031]**	0.135[0.128-0.143]	**0.115[0.111-0.125]**
PSNR	18.98[18.82-19.15]	**22.82[22.61-22.86]**	15.67[15.43-15.88]	**16.9[16.7-17.1]**
SSIM	0.55[0.54-0.56]	**0.63[0.62-0.64]**	0.17[0.16-0.18]	**0.19[0.18-0.20]**

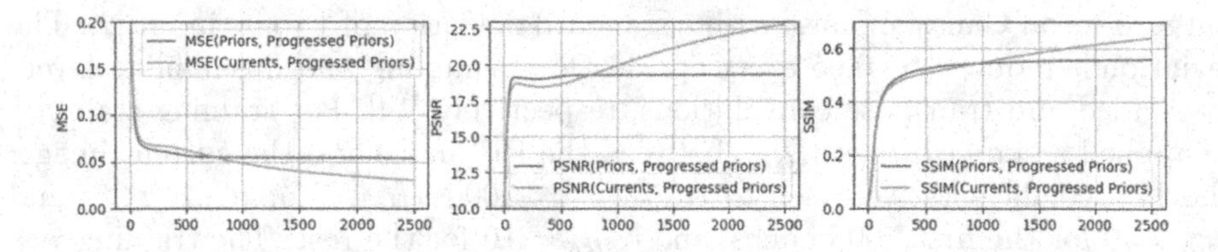

Fig. 3. Image similarity metrics. The X-axis is the epoch number.

3.1 Metrics and Results

To our knowledge, prior works either handle different modalities rather than MGs, or do not tackle the progression task; see Sect. 1. Hence, no comparison to prior works is given. We instead evaluate our results using common quantitative metrics, radiologist evaluation, and visual results. A discussion is given in Sect. 4. We used the following metrics to evaluate our image quality:

Image Similarity. We applied MSE, PSNR, and SSIM metrics [4]. As expected, the similarity between the PP images and the current images increased over the course of training (see Fig. 3). Also, we expected the similarity of the PP images to the prior images to only be slightly affected; see Table 1.

Disease Simulation Quality. We measure the similarity between the simulated disease and the true disease detected in current image using dice metric between the segmentation maps of the current image and either the prior or PP images, as predicted by our pre-trained segmentation model; see Table 1.

Radiologist Evaluation. We presented an experienced radiologist a small sampled dataset of 20 diverse patients including shuffled prior and either current or PP images. The radiologist was asked to classify each MG as real or fake, and decide whether to send the patient for a biopsy (suspicious) or not. We show a similar real/fake statistics when comparing the synthetic PP images to the real prior and current images, and a higher percentage of suspicious images among the PP images than the prior images. This implies that our PP images are realistic MGs that emphasize the subtle disease telltale signs in the prior images. The radiologist was not able to notice any synthetic hints in the lower-resolution images. Hence, only his evaluation on the patch progression is given; see Table 1.

Qualitative Results. See sample results in Fig. 4.

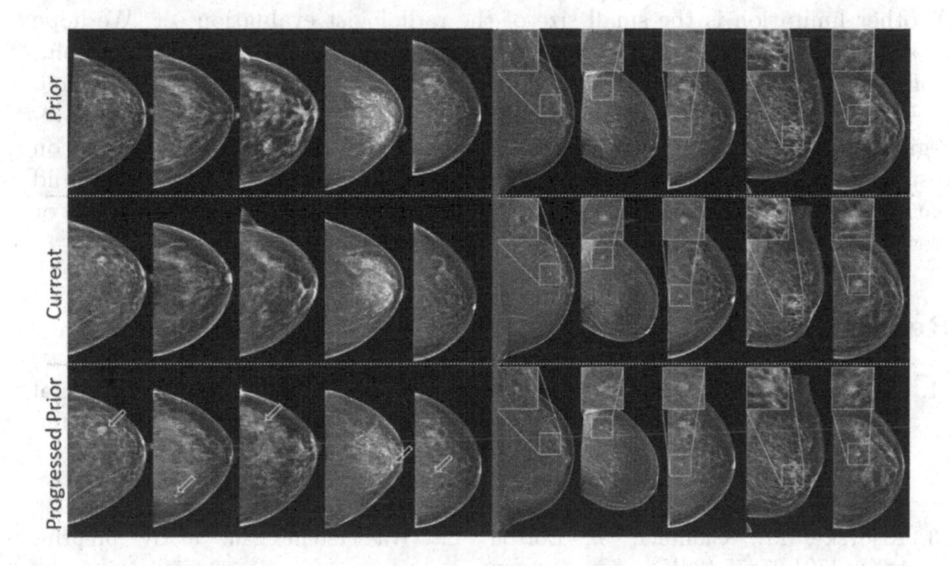

Fig. 4. Qualitative results. (Left): Full image progression, (Right) Patches progression.

4 Discussion, Limitations, and Future Work

In this work we demonstrate the ability to simulate disease progression and generate a new and realistic progressed prior image that: (i) resembles the original prior image, and (ii) includes lesions similar to that of the current image, which were possibly undetected in the prior image. Our novel architecture is trained in a semi-supervised manner on longitudinal data of pairs. We are able to cope with the difficult challenges presented by MG images, including misalignment and the lack of a large annotated longitudinal dataset. We show a dice score increase of 10–15% in both of our tasks, when using our PP images instead of the original priors for lesion segmentation. This points to our model's ability to generate tumors and plant them in the correct position with no supervision. In the radiologist evaluation, an increase of 20% in suspicious biopsy classification means fewer missed breast cancer cases. Thus, our model brings us closer to offering more treatment options, e.g., less radical surgery. It also has the potential to address the problem of over-treatment. Furthermore, Fig. 3 indicates that our PP images still resemble the original prior images, while becoming more similar to the current images as the training progresses. The above results imply that our images indeed look real and satisfy our predefined goals.

Our research was done using a small dataset of longitudinal studies. However, our semi-supervised method enables us to enrich the dataset. The main limitation of this work is the relatively small resolution of the generated full images, as compared to the original 4000×3000 typical size of an MG image. We mitigate this problem by progressing, in full resolution, a suspicious patch. Another limitation is the small size of the radiologist evaluation set. We hope to accumulate more data that will help generate full images in higher resolution. In future work we aim to generate several time stamps of DP, to help us better understand the disease development process, and to improve the pre-registration of MGs using non-parametric registration methods. Also, we plan on testing more organs and modalities, and also incorporating medical history and patient habits (e.g., smoking) to simulate a progression depending on actions or treatment type.

References

1. Abdelhafiz, D., Bi, J., Ammar, R., Yang, C., Nabavi, S.: Convolutional neural network for automated mass segmentation in mammography. BMC Bioinform. **21**(1), 1–19 (2020)
2. Arjovsky, M., Bottou, L.: Towards principled methods for training generative adversarial networks. arXiv preprint arXiv:1701.04862 (2017)
3. Arjovsky, M., Chintala, S., Bottou, L.: Wasserstein gan. arXiv preprint arXiv:1701.07875 (2017)
4. Borji, A.: Pros and cons of gan evaluation measures. Comput. Vis. Image Underst. **179**, 41–65 (2019)
5. Chen, Y.C., et al.: Facelet-bank for fast portrait manipulation. In: Proceedings of the IEEE Conference on Computer Vision and Pattern Recognition, pp. 3541–3549 (2018)
6. Drozdzal, M., Vorontsov, E., Chartrand, G., Kadoury, S., Pal, C.: The importance of skip connections in biomedical image segmentation. In: Carneiro, G., et al. (eds.) LABELS/DLMIA -2016. LNCS, vol. 10008, pp. 179–187. Springer, Cham (2016). https://doi.org/10.1007/978-3-319-46976-8_19
7. Goodfellow, I., et al.: Generative adversarial nets. In: Ghahramani, Z., Welling, M., Cortes, C., Lawrence, N., Weinberger, K.Q. (eds.) Advances in Neural Information Processing Systems, vol. 27, pp. 2672–2680. Curran Associates, Inc. (2014). https://proceedings.neurips.cc/paper/2014/file/5ca3e9b122f61f8f06494c97b1afccf3-Paper.pdf

8. Gulrajani, I., Ahmed, F., Arjovsky, M., Dumoulin, V., Courville, A.: Improved training of wasserstein gans. arXiv preprint arXiv:1704.00028 (2017)
9. IBM Research, H.: Fusemedml: https://github.com/ibm/fuse-med-ml (2021). https://doi.org/10.5281/ZENODO.5146491, https://zenodo.org/record/5146491
10. Korkinof, D., Rijken, T., O'Neill, M., Yearsley, J., Harvey, H., Glocker, B.: High-resolution mammogram synthesis using progressive generative adversarial networks. arXiv preprint arXiv:1807.03401 (2018)
11. Lee, H., Kim, S.T., Lee, J.-H., Ro, Y.M.: Realistic breast mass generation through BIRADS category. In: Shen, D., et al. (eds.) MICCAI 2019. LNCS, vol. 11769, pp. 703–711. Springer, Cham (2019). https://doi.org/10.1007/978-3-030-32226-7_78
12. Mirza, M., Osindero, S.: Conditional generative adversarial nets. arXiv preprint arXiv:1411.1784 (2014)
13. Monticciolo, D., Helvie, M., Hendrick, R.E.: Current issues in the overdiagnosis and overtreatment of breast cancer. Am. J. Roentgenol. **210**, 1–7 (2017). https://doi.org/10.2214/AJR.17.18629
14. Radford, A., Metz, L., Chintala, S.: Unsupervised representation learning with deep convolutional generative adversarial networks. arXiv preprint arXiv:1511.06434 (2015)
15. Ravi, D., Alexander, D.C., Oxtoby, N.P.: Degenerative adversarial NeuroImage nets: generating images that mimic disease progression. In: Shen, D., et al. (eds.) MICCAI 2019. LNCS, vol. 11766, pp. 164–172. Springer, Cham (2019). https://doi.org/10.1007/978-3-030-32248-9_19
16. Ronneberger, O., Fischer, P., Brox, T.: U-Net: convolutional networks for biomedical image segmentation. In: Navab, N., Hornegger, J., Wells, W.M., Frangi, A.F. (eds.) MICCAI 2015. LNCS, vol. 9351, pp. 234–241. Springer, Cham (2015). https://doi.org/10.1007/978-3-319-24574-4_28
17. Sathyan, A., Martis, D., Cohen, K.: Mass and calcification detection from digital mammograms using unets. In: 2020 7th International Conference on Soft Computing & Machine Intelligence (ISCMI), pp. 229–232. IEEE (2020)
18. Shen, T., Gou, C., Wang, F.Y., He, Z., Chen, W.: Learning from adversarial medical images for x-ray breast mass segmentation. Comput. Methods Progr. Biomed. **180**, 105012 (2019)
19. Sun, H., et al.: Aunet: attention-guided dense-upsampling networks for breast mass segmentation in whole mammograms. Phys. Med. Biol. **65**(5), 055005 (2020)
20. Upchurch, P., et al.: Deep feature interpolation for image content changes. In: Proceedings of the IEEE Conference on Computer Vision and Pattern Recognition, pp. 7064–7073 (2017)
21. Wegmayr, V., Hörold, M., Buhmann, J.M.: Generative aging of brain MR-images and prediction of Alzheimer progression. In: Fink, G.A., Frintrop, S., Jiang, X. (eds.) DAGM GCPR 2019. LNCS, vol. 11824, pp. 247–260. Springer, Cham (2019). https://doi.org/10.1007/978-3-030-33676-9_17
22. Zhao, Y., Ma, B., Jiang, P., Zeng, D., Wang, X., Li, S.: Prediction of Alzheimer's disease progression with multi-information generative adversarial network. IEEE J. Biomed. Health Inform. **25**(3), 711–719 (2020). https://doi.org/10.1109/JBHI.2020.3006925

Synth-by-Reg (SbR): Contrastive Learning for Synthesis-Based Registration of Paired Images

Adrià Casamitjana[1](✉), Matteo Mancini[2,3,4], and Juan Eugenio Iglesias[1,5,6]

[1] Center for Medical Image Computing, University College London, London, UK
a.casamitjana@ucl.ac.uk
[2] Department of Neuroscience, University of Sussex, Brighton, UK
[3] NeuroPoly Lab, Polytechnique Montreal, Montreal, Canada
[4] UBRIC, Cardiff University, Cardiff, UK
[5] Martinos Center for Biomedical Imaging, MGH and Harvard Medical School, Boston, USA
[6] Computer Science and AI Laboratory, Massachusetts Institute of Technology, Cambridge, USA

Abstract. Nonlinear inter-modality registration is often challenging due to the lack of objective functions that are good proxies for alignment. Here we propose a synthesis-by-registration method to convert this problem into an easier intra-modality task. We introduce a registration loss for weakly supervised image translation between domains that does not require perfectly aligned training data. This loss capitalises on a registration U-Net with frozen weights, to drive a synthesis CNN towards the desired translation. We complement this loss with a structure preserving constraint based on contrastive learning, which prevents blurring and content shifts due to overfitting. We apply this method to the registration of histological sections to MRI slices, a key step in 3D histology reconstruction. Results on two public datasets show improvements over registration based on mutual information (13% reduction in landmark error) and synthesis-based algorithms such as CycleGAN (11% reduction), and are comparable to registration with label supervision. Code and data are publicly available at https://github.com/acasamitjana/SynthByReg.

Keywords: Image synthesis · Inter-modality registration · Deformable registration · Contrastive estimation

1 Introduction

Image registration is a crucial step to spatially relate information from different medical images. Unpaired registration aligns images of different subjects into a common space to perform subsequent analysis (e.g., population studies [13], voxel-based morphometry [4], or multi-atlas segmentation [18,30]). On the other hand, paired registration aligns different images from the same anatomy and

D. Svoboda et al. (Eds.): SASHIMI 2021, LNCS 12965, pp. 44–54, 2021.
https://doi.org/10.1007/978-3-030-87592-3_5

finds application in image guided intervention (e.g., MR-CT in the prostate [15]); patient follow-up (e.g., pre- and post-operative scans [20]); or longitudinal [29] and multimodal studies (e.g., 3D histology reconstruction with MRI [27]).

Registration is often cast as an optimisation problem where a source image is deformed towards a target image such that it maximises a similarity metric of choice. Classical registration methods solve this problem independently for every pair of images with standard iterative optimisers [32]. Modern learning approaches predict a deformation directly from a pair of images using a convolutional neural network (CNN). Supervised learning methods use ground truth deformation fields in training, either synthetic [31] or derived from manual segmentations [7]. These have been superseded by unsupervised methods, in which CNNs are trained to optimise metrics like those used in classical registration, e.g., sum of squared differences (SSD) or local normalised cross-correlation (LNCC) [6,34], without wasting capacity in regions without salient features.

Widespread similarity functions like SSD or LNCC are well suited for intra-modality registration problems. However, the difficulty of designing accurate similarity functions across modalities hampers inter-modality registration. Mutual information (MI) is often used [21] but with unsatisfactory results in the non-linear case, due to the excessive flexibility of the model [17]. Other metrics used in inter-modality registration are the Modality-Independent Neighbourhood Descriptor (MIND, [14], based on local patch similarities) or adversarial losses measuring whether two images are well aligned or not [12]. MIND is sensitive to initial alignment, bias field or rotations depending on the neighbourhood size, while adversarial losses are prone to missing local correspondences.

An alternative to inter-modality registrationis to convert the problem into an intra-modality task using a registration-by-synthesis framework: image-to-image (I2I) translation is first used to synthesise new source images with the target contrast, and then intra-modality registration (which is more accurate) is performed in the target domain. With accurate image synthesis, the errors introduced by the translation are outweighed by the improvement in registration [17]. In unsupervised synthesis, cycle-consistent generative adversarial networks (CycleGAN) can be used [33,36], but they lack structural consistency across views and may generate artefacts due to overfitting (e.g., flip contrast or even deform images). To mitigate this issue, additional losses between the original and synthetic images have been proposed, e.g., segmentation losses [16] or inter-modality similarities between the original and synthetic scans (e.g., MIND [37] or MI [35]).

Beyond CycleGAN, other approaches have attempted to enforce geometry consistency between the original and synthetic images via specific architectures or training schemes. An I2I translation model that explicitly learns to disentangle domain-invariant (i.e., content) from domain specific features (i.e., appearance) was proposed in [28]; the latent content features can then be used to train a registration network. More recently, a novel training scheme that forces the translation and registration steps to be commutative (thus discouraging deformation at synthesis) has been presented [2]. Nonetheless, GAN-based approaches are challenging to train, with well-known problems (e.g., vanishing gradients,

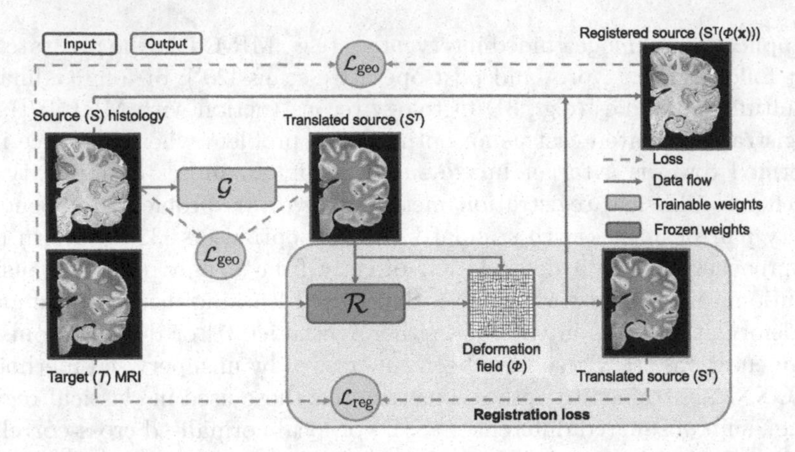

Fig. 1. Overview of proposed pipeline, using histology and MRI as source and target contrasts, respectively.

instability [3]) and an increasing number of losses and hyperparameters. In this work, we turn the registration-by-synthesis framework around into a synthesis-by-registration (SbR) approach, where a registration network trained on the target domain (and frozen weights) is used in the loss for training an I2I network. This allows us to greatly simplify the objective function and avoid potentially unstable adversarial training. Moreover, we use contrastive learning at the patch level to ensure geometric consistency. The SbR model outputs both the translated image and the deformation field. The contribution of this work is threefold: *(i)* we develop a novel registration loss for paired I2I translation; *(ii)* we adapt the contrastive PatchNCE loss [26] for image registration as a geometry-preserving constraint; and *(iii)* we combine *(i)* and *(ii)* into an unsupervised SbR framework for inter-modality registration that does not require multiple encoders/decoders and therefore has low GPU memory requirements.

2 Methods

2.1 Overview

Let us consider two misaligned 2D images of the same anatomy (e.g., a histological section and a corresponding MRI plane): the source $S(\boldsymbol{x})$ and target $T(\boldsymbol{x})$; $\boldsymbol{x}$ represents spatial location. We further assume the availability of an intra-modality registration CNN $\mathcal{R}$ with weights $\boldsymbol{\theta}_{\mathcal{R}}$, which predicts a deformation field Φ from two images of the target modality: $\Phi = \mathcal{R}(T, T'; \boldsymbol{\theta}_{\mathcal{R}})$, such that $T(\boldsymbol{x}) \approx T'(\Phi(\boldsymbol{x}))$. We also define an I2I translation CNN $\mathcal{G}$ (with weights $\boldsymbol{\theta}_{\mathcal{G}}$) from contrast S to T that regresses the image intensities: $S^T = \mathcal{G}(S; \boldsymbol{\theta}_{\mathcal{G}})$, such that S^T resembles the anatomy in S, had it been acquired with modality T. The crucial observation is that, if S^T is well synthesised, and $\Phi = \mathcal{R}(T, S^T; \boldsymbol{\theta}_{\mathcal{R}})$, then $T(\boldsymbol{x}) \approx S^T(\Phi(\boldsymbol{x}))$. Specifically, we propose the following loss (Fig. 1):

$$\mathcal{L}(\boldsymbol{\theta}_{\mathcal{G}}) = \mathcal{L}_{\text{reg}}(S, T; \boldsymbol{\theta}_{\mathcal{G}}, \boldsymbol{\theta}_{\mathcal{R}}) + \lambda_{\text{geo}} \mathcal{L}_{\text{geo}}(S, T; \boldsymbol{\theta}_{\mathcal{G}}, \boldsymbol{\theta}_{\mathcal{R}}), \tag{1}$$

where $\mathcal{L}_{\text{reg}}$ is a "registration loss" measuring the similarity of T and (the deformed) S^T, $\mathcal{L}_{\text{geo}}$ is a geometric consistency loss that ensures that the contents of S and S^T are aligned, and λ_{geo} is a relative weight. A key implicit assumption of this framework is that, because the images are paired, there exists a spatial transform Φ that aligns S and T well, such that the synthesis does not need to shift or blur boundaries to minimise the error; the geometric consistency loss further discourages such mistakes. Crucially, the loss in Eq. 1 does not depend on $\boldsymbol{\theta}_{\mathcal{R}}$: the registration CNN is trained on the target domain and its weights are frozen, such that gradients will backpropagate through these layers to improve the synthesis. This asymmetric scheme enables us to avoid using a distribution matching loss (e.g., CycleGAN) that may produce hallucination artefacts [9].

2.2 Intra-modality Registration Network

One of the key points in SbR is the differentiable registration method used to train the image synthesis model. We use a U-Net [8] model (as in [10]) that learns a diffeomorphic mapping between images from the same modality. The model is trained on pairs of images from the target domain and outputs a stationary velocity field (SVF), $\boldsymbol{\psi}$, at half the input resolution. Then, a scaling and squaring approach is used to integrate $\boldsymbol{\psi}$ into a half-resolution deformation field, which is linearly upsampled to obtain the final deformation $\Phi(\boldsymbol{x})$. Training uses LNCC as image similarity term, and the norm of the gradient of the SVF as regulariser:

$$\mathcal{L}_{\mathcal{R}}(T, T'; \boldsymbol{\theta}_{\mathcal{R}}) = \frac{1}{|\Omega|} \sum_{\boldsymbol{x} \in \Omega} LNCC[T(\boldsymbol{x}), T'(\Phi(\boldsymbol{x}; \boldsymbol{\theta}_{\mathcal{R}}))] + 2\lambda_{\mathcal{R}} \|\nabla \boldsymbol{\psi}(\boldsymbol{x}; \boldsymbol{\theta}_{\mathcal{R}})\|_2, \quad (2)$$

where Ω is the discrete image domain, and $\lambda_{\mathcal{R}}$ is a relative weight. Since the goal is to learn registration of images with approximately the same anatomy, we train the CNN with pairs of images that are similar to each other – specifically, within 3 neighbours in the image stack. In order to prevent overfitting, which may be problematic due to the relatively limited number of combinations of pairs, we use random spatial transformations for data augmentation at each iteration in the source and target images, including small random similarity transforms and smooth nonlinear deformations. Once this CNN has been trained, its weights $\boldsymbol{\theta}_{\mathcal{R}}$ are frozen during training of the rest of layers in our framework.

2.3 Image-to-Image Translation Using a Registration Loss

The modality translation is performed by a generator network, $\mathcal{G}$, with a similar architecture to [26] and trained using a combination of two losses: $\mathcal{L}_{\text{reg}}$ and $\mathcal{L}_{\text{geo}}$. The first component $\mathcal{L}_{\text{reg}}$ is the registration loss between the target and the translated, deformed source. In Sect. 2.2 above, we used the LNCC metric, which is known to work well in learning-based, intra-modality registration registration of most modalities, and can handle bias field in MRI [6]. However, in I2I we need to explicitly penalise absolute intensity differences, since encouraging local correlation is not enough to optimise the synthesis. For this purpose, we use the

ℓ_1-norm, which has been widely used in the synthesis literature, and which is more robust than ℓ_2 against violations of the assumption that the anatomy is perfectly paired in the source and target images. The registration loss is:

$$\mathcal{L}_{\text{reg}}(S, T; \boldsymbol{\theta}_{\mathcal{G}}, \boldsymbol{\theta}_{\mathcal{R}}) = \frac{1}{|\Omega|} \sum_{\boldsymbol{x} \in \Omega} \|T(\boldsymbol{x}) - S^T(\Phi(\boldsymbol{x}; \boldsymbol{\theta}_{\mathcal{R}}); \boldsymbol{\theta}_{\mathcal{G}})\|_1. \quad (3)$$

The second component of the loss $\mathcal{L}_{\text{geo}}$ seeks to enforce geometric consistency in the synthesis and is based on noise contrastive estimation (PatchNCE [26]). The idea behind PatchNCE is to maximise a lower bound on the MI between the pre- and post-synthesis images at the patch level. For this purpose, we define a "query" image q (e.g., S^T) and a "reference" image r (e.g., S), from which we extract patch descriptors from the stack of features computed by the encoding branch of the I2I CNN, $\mathcal{G}$. These descriptors are the output of L layers of interest, including: the input image, the downsampling convolutional layers and the first and last ResNet blocks. Specifically, we extract sets of features $\boldsymbol{f}_l$ at the layers of interest $l = 1, \dots, L$ and N random locations $\boldsymbol{x}_{l,n}$ per layer, i.e., $\{\boldsymbol{f}_l(\boldsymbol{x}_{l,n})\}_{l=1,\dots,L; n=1,\dots,N}$ (in practice, a tissue mask is used when drawing $\boldsymbol{x}_{l,n}$ in order not to sample the background). Each of these $\boldsymbol{f}_l$ encodes different image features (with different number of channels), from different neighbourhoods (patches), and at different resolution levels.

Given these descriptors, the contrastive loss builds on the principle that $\boldsymbol{f}_l^q(\boldsymbol{x}_{l,n})$ (for the query) and $\boldsymbol{f}_l^r(\boldsymbol{x}_{l,n'})$ (for the reference) should be similar for $n = n'$ and dissimilar for $n \neq n'$. Rather than using the descriptors $\boldsymbol{f}$ directly, we follow in [26] and run them through two-layer perceptrons (which are different for the descriptors in every layer l, since they have different resolutions), followed by unit-norm normalisation layers. This yields a new representation $\{\boldsymbol{z}_{l,n}\}_{l=1,\dots,L; n=1,\dots,N}$, with:

$$\boldsymbol{z}_{l,n} = \mathcal{Q}_l(\boldsymbol{f}_l(\boldsymbol{x}_{l,n}); \boldsymbol{\theta}_z), \quad (4)$$

where $\boldsymbol{\theta}_z$ groups the parameters of these representation layers. Given $\boldsymbol{z}$, the contrastive PatchNCE loss is given by a softmax function of cosine similarities:

$$\mathcal{L}_{\text{PatchNCE}}(q, r; \boldsymbol{\theta}_z, \tau) = -\frac{1}{N} \sum_{n=1}^{N} \sum_{l=1}^{L} \log \left(\frac{\exp(\boldsymbol{z}_{l,n}^q \cdot \boldsymbol{z}_{l,n}^r / \tau)}{\sum_{n'=1}^{N} \exp(\boldsymbol{z}_{l,n}^q \cdot \boldsymbol{z}_{l,n'}^r / \tau)} \right), \quad (5)$$

where τ is a temperature parameter and $(\cdot)$ is the dot product. It can be shown that the lower bound on the MI becomes tighter with increasing N [25].

In practice, we use two PatchNCE losses: one between the source and translated images; and another between the registered and target images:

$$\begin{aligned}\mathcal{L}_{\text{geo}}(S, T; \boldsymbol{\theta}_{\mathcal{G}}, \boldsymbol{\theta}_{\mathcal{R}}, \boldsymbol{\theta}_z, \tau) = &\mathcal{L}_{\text{PatchNCE}}(S^T(\boldsymbol{x}; \boldsymbol{\theta}_{\mathcal{G}}), S(\boldsymbol{x}); \theta_z, \tau) \\ &+ \mathcal{L}_{\text{PatchNCE}}(S^T(\Phi(\boldsymbol{x}; \boldsymbol{\theta}_{\mathcal{R}}); \boldsymbol{\theta}_{\mathcal{G}}), T(\boldsymbol{x}); \theta_z, \tau). \quad (6)\end{aligned}$$

Combining the registration and geometric consistency losses in Eqs. 3 and 7 yields the final loss for our meta-architecture:

$$\begin{aligned}\mathcal{L}(\boldsymbol{\theta}_{\mathcal{G}}, \boldsymbol{\theta}_z) = & \frac{1}{|\Omega|} \sum_{\boldsymbol{x} \in \Omega} \|T(\boldsymbol{x}) - S^T(\Phi(\boldsymbol{x}; \boldsymbol{\theta}_{\mathcal{R}}); \boldsymbol{\theta}_{\mathcal{G}})\|_1 \\ & + \lambda_{\text{geo}} \mathcal{L}_{\text{PatchNCE}}(S^T(\boldsymbol{x}; \boldsymbol{\theta}_{\mathcal{G}}), S(\boldsymbol{x}); \theta_z, \tau) \\ & + \lambda_{\text{geo}} \mathcal{L}_{\text{PatchNCE}}(S^T(\Phi(\boldsymbol{x}; \boldsymbol{\theta}_{\mathcal{R}}); \boldsymbol{\theta}_{\mathcal{G}}), T(\boldsymbol{x}); \theta_z, \tau), \end{aligned} \quad (7)$$

which we optimise with respect to $\boldsymbol{\theta}_{\mathcal{G}}$ and $\boldsymbol{\theta}_z$ – since τ is a fixed hyperparameter and $\boldsymbol{\theta}_{\mathcal{R}}$ is frozen, as explained above.

3 Experiments and Results

3.1 Data

We validate the presented methodology in the context of 3D histology reconstruction via registration to a reference MRI volume. We use two publicly available datasets with histological sections and an *ex vivo* 3D MRI of the same subject. A 3D similarity transform between the stack of histological sections and the MRI volume was used to align images from both domains [24]. The MRI volume was then resampled into the space of histological stack, which yields a set of paired images to register: histological sections and corresponding MRI resampled planes. The two datasets are:

- **Allen Human Brain Atlas** [11]: this dataset includes 93 sections with manual delineations of hundreds of brain structures, which we grouped into four coarse tissue classes: cerebral white matter (WM), cerebral grey matter (GM), cerebellar white matter (WMc), and cerebellar grey matter (GMc). An *ex vivo* MRI is available, which was segmented into the same four tissue classes with SPM [5]. In addition, J.E.I. manually annotated 13.8±4.4 pairs of matching landmarks in the histological sections and corresponding resampled MRI planes, uniformly distributed across all spatial locations.
- **BigBrain Initiative** [1]: we considered one every 20 sections, i.e., one section every 0.4 mm (344 sections in total). As in the previous dataset, an *ex vivo* MRI is available and J.E.I. manually annotated 11.6 ± 1.7 landmark pairs in the histological sections and corresponding MRI planes. No segmentations are available for this dataset.

3.2 Experimental Setup

In our experiments, we register each histological section to the corresponding (resampled) MRI slice. For quantitative evaluation, we report the average root-mean-squared landmark error (both datasets) and the Dice score on brain tissue classes (only for the Allen dataset).

Our proposed method, **SbR**, was trained with the following hyper-parameters: $\lambda_{\text{geo}} = 0.02$, $\tau = 0.05$ and $\lambda_{\mathcal{R}} = 1$, which were set from a subset of

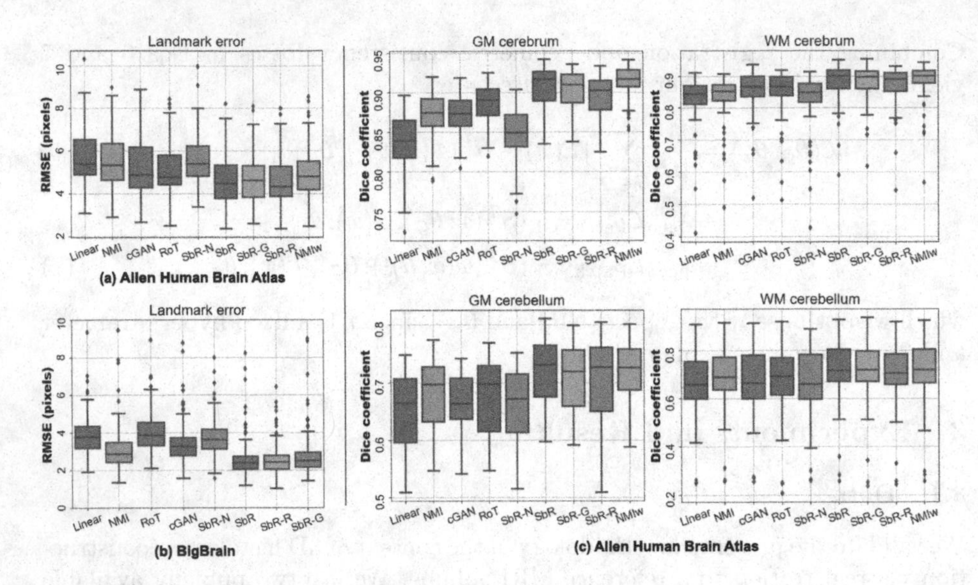

Fig. 2. Landmark mean squared error on the Allen human brain atlas dataset (a) and the BigBrain dataset (b). Dice score coefficient for the Allen dataset is shown in (c).

the Allen dataset and used elsewhere. We also tested three other configurations of our method: an ablated version without the structure preserving constraint, i.e., $\lambda_{\text{geo}} = 0$ (**SbR-N**); fine-tuning the result of **SbR** by unfreezing the registration parameters (**SbR-R**); and an extension (**SbR-G**) that includes an LSGAN loss [22] with a PatchGAN discriminator [19] to discriminate between synthesised and target images (S^T and T). **SbR-G** enables us to assess the potential benefits of adding a distribution matching loss in training.

In addition, we compare our method against a number of other methods, to test differences against: standard registration metrics, other synthesis-based approaches without specific geometric constraints, and supervision with labels and Dice scores. Specifically, the competing methods are: (*i*) **Linear**, the initial affine registration with NiftyReg [24]; (*ii*) **NMI**, unsupervised training using normalised mutual information (NMI) with 20 bins on the image intensities; (*iii*) **NMIw**, weakly supervised training using NMI and an additional Dice loss [23] on the segmentations; (*iv*) **cGAN**, a CycleGAN [38] approach combined with our registration loss; and (*v*) **RoT**, the state-of-the-art method presented in [2] that consists of alternating the registration and translation steps. All learning-based methods above (including ours) use the same architecture for registration, and also the same nonlinear spatial augmentation scheme (sampling $9 \times 9 \times 2$ from zero-mean Gaussians and upsampling to full resolution)

3.3 Results

The quantitative results are summarised in Fig. 2. The landmark errors show that our proposed method (SbR) outperforms all baseline approaches: 11%, 9% and

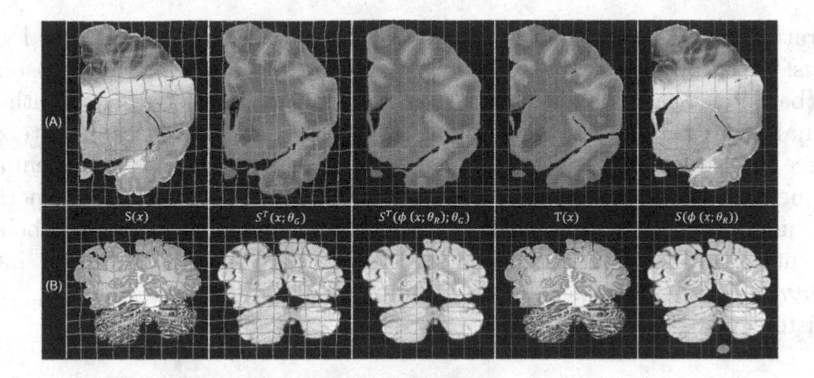

Fig. 3. Image examples from (a) the Allen human brain atlas, and (b) the BigBrain project, with the deformed and rectangular grid overlaid on the source and target spaces, respectively.

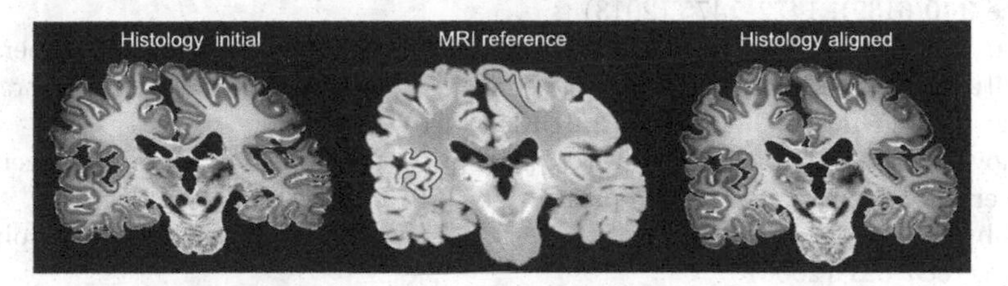

Fig. 4. Section 170 from BigBrain, with cortical boundaries manually traced on the target domain (MRI) and overlaid on the histology, before and after registration.

7% error reduction with respect cGAN, RoT and NMIw in the Allen dataset and 23% and 33% with respect cGAN and RoT in the BigBrain dataset; all improvements are statistically significant ($p < 0.001$) using a Wilcoxon signed-rank test. Interestingly, SbR is able to align tissue masks as well as NMIw, even though segmentations were not used in the training phase. The naive approach (SbR-N) suffers from synthetic artefacts in the generator, which degrades the results - thus highlighting the importance of including structure preserving constraints in the model. The other two extensions of the model, SbR-G and SbR-R, achieve similar performance to the initial configuration, without yielding any statistically significant additional benefits.

In Fig. 3, we show an example of the synthesised and registered images using SbR for each dataset. The method displays robustness against common artefacts, such as: cracks, missing tissue and inhomogeneous staining (in histology), or intensity inhomogeneity (in MRI). Our method is able to accurately register convoluted structures such as the cortex, as seen in Fig. 4.

4 Discussion and Conclusion

We have presented Synth-by-Reg, a synthesis-by-registration framework for inter-modality registration, which we have validated on a histology-to-MRI registration task. The method uses a single I2I translation network trained with a robust registration loss (based on the ℓ_1-norm) and a geometric consistency term (based on contrastive learning). In histology-MRI registration, Synth-by-Reg enables us to avoid using a CycleGAN approach, which often falters in presence of histological artefacts – since it needs to learn to simulate them and subsequently recover from them. Future work will focus on adapting our method to the unpaired scenario, as well as to other imaging modalities. We believe that synthesis-by-registration can be a very useful alternative in difficult inter-modality registration problems when weakly paired data are available, e.g., MRI and histology.

References

1. Amunts, K., et al.: BigBrain: an ultrahigh-resolution 3D human brain model. Science **340**(6139), 1472–1475 (2013)
2. Arar, M., Ginger, Y., Danon, D., Bermano, A.H., Cohen-Or, D.: Unsupervised multi-modal image registration via geometry preserving image-to-image translation. In: CVPR, pp. 13410–13419. IEEE (2020)
3. Arjovsky, M., Bottou, L.: Towards principled methods for training generative adversarial networks. arXiv preprint arXiv:1701.04862 (2017)
4. Ashburner, J., Friston, K.: Voxel-based morphometry-the methods. Neuroimage **11**(6), 805–821 (2000)
5. Ashburner, J., Friston, K.: Unified segmentation. Neuroimage **26**, 839–851 (2005)
6. Balakrishnan, G., Zhao, A., Sabuncu, M.R., Guttag, J., Dalca, A.V.: Voxelmorph: a learning framework for deformable medical image registration. IEEE Trans. Med. Imaging **38**(8), 1788–1800 (2019)
7. Cao, X., et al.: Deformable image registration based on similarity-steered CNN regression. In: Descoteaux, M., Maier-Hein, L., Franz, A., Jannin, P., Collins, D.L., Duchesne, S. (eds.) MICCAI 2017. LNCS, vol. 10433, pp. 300–308. Springer, Cham (2017). https://doi.org/10.1007/978-3-319-66182-7_35
8. Çiçek, Ö., Abdulkadir, A., Lienkamp, S.S., Brox, T., Ronneberger, O.: 3D U-Net: Learning Dense Volumetric Segmentation from Sparse Annotation. In: Ourselin, S., Joskowicz, L., Sabuncu, M.R., Unal, G., Wells, W. (eds.) MICCAI 2016. LNCS, vol. 9901, pp. 424–432. Springer, Cham (2016). https://doi.org/10.1007/978-3-319-46723-8_49
9. Cohen, J.P., Luck, M., Honari, S.: Distribution matching losses can hallucinate features in medical image translation. In: Frangi, A.F., Schnabel, J.A., Davatzikos, C., Alberola-López, C., Fichtinger, G. (eds.) MICCAI 2018. LNCS, vol. 11070, pp. 529–536. Springer, Cham (2018). https://doi.org/10.1007/978-3-030-00928-1_60
10. Dalca, A.V., Balakrishnan, G., Guttag, J., Sabuncu, M.R.: Unsupervised learning for fast probabilistic diffeomorphic registration. In: Frangi, A.F., Schnabel, J.A., Davatzikos, C., Alberola-López, C., Fichtinger, G. (eds.) MICCAI 2018. LNCS, vol. 11070, pp. 729–738. Springer, Cham (2018). https://doi.org/10.1007/978-3-030-00928-1_82
11. Ding, S.L., et al.: Comprehensive cellular-resolution atlas of the adult human brain. J. Comp. Neurol. **524**(16), 3127–3481 (2016)
12. Fan, J., Cao, X., Wang, Q., Yap, P.T., Shen, D.: Adversarial learning for mono-or multi-modal registration. Med. Image Anal. **58**, 101545 (2019)

13. Fonov, V., Evans, A.C., Botteron, K., Almli, C.R., McKinstry, R.C., Collins, D.L.: Unbiased average age-appropriate atlases for pediatric studies. Neuroimage **54**(1), 313–327 (2011)
14. Heinrich, M.P., et al.: MIND: modality independent neighbourhood descriptor for multi-modal deformable registration. Med. Image Anal. **16**(7), 1423–1435 (2012)
15. Hu, Y.: MR to ultrasound registration for image-guided prostate interventions. Med. Image Anal. **16**(3), 687–703 (2012)
16. Huo, Y., Xu, Z., Bao, S., Assad, A., Abramson, R.G., Landman, B.A.: Adversarial synthesis learning enables segmentation without target modality ground truth. In: ISBI, pp. 1217–1220. IEEE (2018)
17. Iglesias, J.E., Konukoglu, E., Zikic, D., Glocker, B., Van Leemput, K., Fischl, B.: Is synthesizing MRI contrast useful for inter-modality analysis? In: Mori, K., Sakuma, I., Sato, Y., Barillot, C., Navab, N. (eds.) MICCAI 2013. LNCS, vol. 8149, pp. 631–638. Springer, Heidelberg (2013). https://doi.org/10.1007/978-3-642-40811-3_79
18. Iglesias, J.E., Sabuncu, M.R.: Multi-atlas segmentation of biomedical images: a survey. Med. Image Anal. **24**(1), 205–219 (2015)
19. Isola, P., Zhu, J.Y., Zhou, T., Efros, A.A.: Image-to-image translation with conditional adversarial networks. In: CVPR, pp. 1125–1134. IEEE (2017)
20. Kwon, D., Niethammer, M., Akbari, H., Bilello, M., Davatzikos, C., Pohl, K.M.: PORTR: pre-operative and post-recurrence brain tumor registration. IEEE Trans. Med. Imaging **33**(3), 651–667 (2013)
21. Maes, F., Vandermeulen, D., Suetens, P.: Medical image registration using mutual information. Proc. IEEE **91**(10), 1699–1722 (2003)
22. Mao, X., Li, Q., Xie, H., Lau, R.Y., Wang, Z., Paul Smolley, S.: Least squares generative adversarial networks. In: CVPR, pp. 2794–2802. IEEE (2017)
23. Milletari, F., Navab, N., Ahmadi, S.A.: V-net: fully convolutional neural networks for volumetric medical image segmentation. In: 3DV Conference, pp. 565–571 (2016)
24. Modat, M., et al.: Fast free-form deformation using graphics processing units. Comput. Methods Programs Biomed. **98**(3), 278–284 (2010)
25. Oord, A.V.D., Li, Y., Vinyals, O.: Representation learning with contrastive predictive coding. arXiv preprint arXiv:1807.03748 (2018)
26. Park, T., Efros, A.A., Zhang, R., Zhu, J.-Y.: Contrastive learning for unpaired image-to-image translation. In: Vedaldi, A., Bischof, H., Brox, T., Frahm, J.-M. (eds.) ECCV 2020. LNCS, vol. 12354, pp. 319–345. Springer, Cham (2020). https://doi.org/10.1007/978-3-030-58545-7_19
27. Pichat, J., Iglesias, J.E., Yousry, T., Ourselin, S., Modat, M.: A survey of methods for 3D histology reconstruction. Med. Image Anal. **46**, 73–105 (2018)
28. Qin, C., Shi, B., Liao, R., Mansi, T., Rueckert, D., Kamen, A.: Unsupervised deformable registration for multi-modal images via disentangled representations. In: Chung, A.C.S., Gee, J.C., Yushkevich, P.A., Bao, S. (eds.) IPMI 2019. LNCS, vol. 11492, pp. 249–261. Springer, Cham (2019). https://doi.org/10.1007/978-3-030-20351-1_19
29. Reuter, M., Schmansky, N.J., Rosas, H.D., Fischl, B.: Within-subject template estimation for unbiased longitudinal image analysis. Neuroimage **61**, 1402–18 (2012)
30. Rohlfing, T., Brandt, R., Menzel, R., Maurer, C.R., Jr.: Evaluation of atlas selection strategies for atlas-based image segmentation with application to confocal microscopy images of bee brains. Neuroimage **21**(4), 1428–1442 (2004)

31. Sokooti, H., de Vos, B., Berendsen, F., Lelieveldt, B.P.F., Išgum, I., Staring, M.: Nonrigid image registration using multi-scale 3D convolutional neural networks. In: Descoteaux, M., Maier-Hein, L., Franz, A., Jannin, P., Collins, D.L., Duchesne, S. (eds.) MICCAI 2017. LNCS, vol. 10433, pp. 232–239. Springer, Cham (2017). https://doi.org/10.1007/978-3-319-66182-7_27
32. Sotiras, A., Davatzikos, C., Paragios, N.: Deformable medical image registration: a survey. IEEE Trans. Med. Imaging **32**(7), 1153–1190 (2013)
33. Tanner, C., Ozdemir, F., Profanter, R., Vishnevsky, V., Konukoglu, E., Goksel, O.: Generative adversarial networks for MR-CT deformable image registration. arXiv preprint arXiv:1807.07349 (2018)
34. de Vos, B.D., Berendsen, F.F., Viergever, M.A., Staring, M., Išgum, I.: End-to-end unsupervised deformable image registration with a convolutional neural network. In: Cardoso, M.J., et al. (eds.) DLMIA/ML-CDS -2017. LNCS, vol. 10553, pp. 204–212. Springer, Cham (2017). https://doi.org/10.1007/978-3-319-67558-9_24
35. Wang, C., Yang, G., Papanastasiou, G., Tsaftaris, S.A., Newby, D.E., Gray, C., et al.: DiCyc: GAN-based deformation invariant cross-domain information fusion for medical image synthesis. Inf. Fus. **67**, 147–160 (2021)
36. Wei, D., et al.: Synthesis and inpainting-based MR-CT registration for image-guided thermal ablation of liver tumors. In: Shen, D., et al. (eds.) MICCAI 2019. LNCS, vol. 11768, pp. 512–520. Springer, Cham (2019). https://doi.org/10.1007/978-3-030-32254-0_57
37. Xu, Z., et al.: Adversarial uni- and multi-modal stream networks for multimodal image registration. In: Martel, A.L., et al. (eds.) MICCAI 2020. LNCS, vol. 12263, pp. 222–232. Springer, Cham (2020). https://doi.org/10.1007/978-3-030-59716-0_22
38. Zhu, J.Y., Park, T., Isola, P., Efros, A.A.: Unpaired image-to-image translation using cycle-consistent adversarial networks. In: CVPR, pp. 2223–2232. IEEE (2017)

Learning-Based Template Synthesis for Groupwise Image Registration

Ziyi He[✉] and Albert C. S. Chung

Department of Computer Science and Engineering, The Hong Kong University of Science and Technology, Kowloon, Hong Kong
zheaj@connect.ust.hk, achung@cse.ust.hk

Abstract. Groupwise image registration (GIR), as a fundamental task during medical image processing, aims to transform a group of images into a common space simultaneously. Most GIR methods consist of creating one template image and registering all group subjects to the template space. This paper presents a novel learning-based template synthesis method that can produce sharper and unbiased template images in shorter runtime. The method is based on the generative adversarial network (GAN) scheme, which is expected to generate authentic images. Besides GAN, we use a registration network to warp images and an auxiliary segmentor to take advantage of semantic information and improve the quality of generated images. First, the label mask will be combined with medical images as the input of the discriminator. Second, we calculate the entropy of pixel-wise label probability distribution as a new loss term. We compare our method with three baseline methods on 2D and 3D brain MR images in the experiments. We evaluate the performance using metrics corresponding to accuracy, smoothness, and unbiasedness. Results illustrate that our method outperforms the baseline methods and can achieve the overall state-of-the-art performance.

Keywords: Template synthesis · Groupwise image registration · Generative adversarial networks

1 Introduction

Image registration is the process of estimating the optimal spatial transformation between images with matched content. When analyzing medical images acquired from different scanning planes, at different times, or using different modalities, accurate registration helps researchers look into tiny differences or changes [2,17]. The changes can be between images for both intra-subject changes in longitudinal studies and the inter-subject variability in cross-sectional studies [11].

As imaging technology becomes more efficient and advanced, the scale of image data increases exponentially. Therefore, the need to process a bunch of medical images arises. Unlike pairwise registration, groupwise image registration (GIR) is proposed to find deformation fields for a group of images and warp

D. Svoboda et al. (Eds.): SASHIMI 2021, LNCS 12965, pp. 55–66, 2021.
https://doi.org/10.1007/978-3-030-87592-3_6

them into a common center. GIR has already become an essential task in medical image analysis and clinical studies like fusing images from different modalities or subjects, quantifying the change of organs during the treatment, and so on. Most of the existing GIR methods are developed based on pairwise registration methods, which first construct a template image and then register each subject to the template. The template can be iteratively updated. Non-learning based template construction algorithms include group mean [15], sharp mean [24], and some dynamic weighting strategy [19]. More recently, Che et al. [8] proposed to combine principal component analysis with deep learning-based registration. [9] presented a probabilistic deformation model which takes the target group as the training set for unconditional large group template construction. Besides, some researchers apply different data structures such as tree-like or graph-like structure to perform GIR in a hierarchical way like [10,22]. Though most methods rely on a given or generated template, some researchers are also exploring template-free settings [1,13] which benefit from more constraints within the group, to deform group subjects without any reference images.

In this paper, we perform GIR with templates because the explicit template image can work as the registration target and the representative of a group compared with template-free methods. We propose to synthesize the template from the group using the deep learning network. Conventional image synthesis methods like dictionary learning [23] deal with selected and handcrafted features that cannot represent complex visual information [26]. Recently, there are learning-based research works proposed for medical image synthesis. Dong et al. [18] were among the first to apply a 3D fully convolutional neural network to estimate CT images from MR images supervisedly. Later, the thrive of generative adversarial networks (GANs) [12] has catalyzed the development of cutting-edge image synthesis works [4,5,7]. Further, Zhang et al. [28] took the global contextual information into consideration and introduced a sketch prior constraint to guide the image generation. Yu et al. [27] proposed an edge-aware GAN to keep textual structures of the generated images.

To further improve the quality of the synthetic template, we utilize the segmentation information of medical images to improve the semantic reliability and sharpness of the generated images. Some previous work [6,29] have introduced a segmentor into the cross-modality image translation task using GAN. However, these methods compare the segmentation masks between the generated image and the corresponding real image which we do not have, so they cannot be directly applied to the groupwise registration framework.

To conclude, we propose a novel learning-based template synthesis method for groupwise image registration. The contributions can be summarized as:

- We are the first to use segmentation assisted GAN to quickly generate authentic medical template images for given groups;
- We introduce the auxiliary segmentation module into the framework to utilize semantic and texture information, from which we design an entropy loss term to improve the quality of generated templates;

- Experiments conducted on MR image slices and volumes illustrate that our framework outperforms baseline methods in the aspect of registration accuracy, displacement field smoothness, template unbiasedness, and runtime.

2 Method

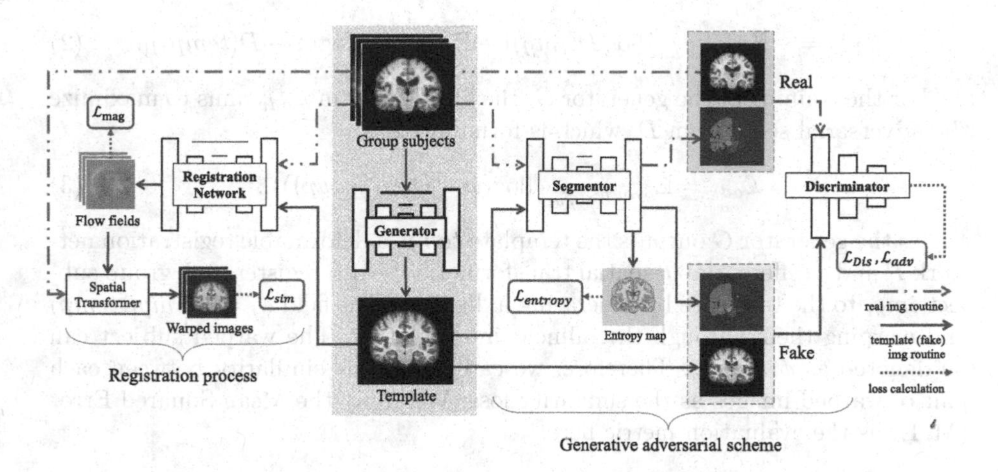

Fig. 1. Framework of the proposed method. The central module shows that the group subjects will be fed into the generator G, and G generates the template. The segmentor S predicts the label mask, which will be used in addition to MR images as the discern target for the discriminator D. We also construct an entropy map using the possibility distribution generated during the segmentation, which will improve the sharpness. On the left side, the registration network R will take the group subjects and the template to predict the displacement fields and warp the images.

To quickly synthesis an authentic template image, we adopt the GAN scheme, which involves a generator to create images and a discriminator to discern whether the input looks authentic or not. A deformable registration network is applied to predict displacement fields for warping group subjects to the template. In addition to the image-level distinction, we add the segmentation mask as the second discriminant to utilize semantic and contextual information. We derive an entropy loss term from the segmentation mask that improves the template quality (Fig. 1).

2.1 Generative Adversarial Networks

The generator G takes N group subjects $group = \{img_1, img_2, ..., img_N\}$ as the input and outputs the template image $temp = G(group)$ of the same size of group subjects. The discriminator D aims to recognize whether the input img is authentic or fake (from the generator). The minimax objective can be formulated as:

$$\min_{G} \max_{D} \quad \mathbb{E}_{img\sim\mathbb{P}_{data}}[\log D(img)] + \mathbb{E}_{group\sim\mathbb{P}^{N}_{data}}[\log(1 - D(G(group))]. \quad (1)$$

The objective of the discriminator D is to maximize the score of real images and minimize the score of fake images. Thus, the loss function of training D is :

$$\mathcal{L}_{Dis} = -\,\mathbb{E}_{img\sim\mathbb{P}_{data}}[\log D(img)] - \mathbb{E}_{temp\sim\mathbb{P}_{G}}[\log(1 - D(temp))]. \quad (2)$$

For the training of the generator G, the first loss term $\mathcal{L}_{adv}$ aims to maximize the adversarial score from D, which is formulated as:

$$\mathcal{L}_{adv} = \mathbb{E}_{group\sim\mathbb{P}^{N}_{data}}[\log(1 - D(G(group)))]. \quad (3)$$

As the generator G outputs the template *temp*, a deformable registration network R and a differentiable spatial transformer STN [14] register each group subject img_i to the template by generating a displacement field $\phi_i = R(img_i, temp)$ and warping them through the bilinear interpolation. The warped subject can be denoted as $\phi_i \circ img_i$. Therefore, we can utilize the similarity between each pair of warped images as the similarity loss. We apply the Mean Squared Error (MSE) as the evaluation metric here:

$$\mathcal{L}_{sim} = \frac{1}{N}\sum_{i}^{N}(temp - \phi_i \circ img_i)^2, \mathcal{L}_{mag} = \frac{1}{N}||\phi_i||_2^2. \quad (4)$$

Besides, to force the template to be close to the center of the latent center space and to penalize large displacements, we add a loss term $\mathcal{L}_{mag}$ to restraint the magnitude of the displacement fields.

2.2 Auxiliary Segmentor

The basic GAN scheme mentioned above that generates authentic MR images at the image element (pixel/voxel) level may still create artifacts like blurring and posterizing. To further improve the reliability of the template image, we train an auxiliary segmentation network S separately and utilize the masks' semantic and context information to optimize the basic GAN + Registration framework.

First of all, we let the discriminator D take the segmentation mask $seg = S(img)$ of the input image into account. Hence D discerns the concatenation of the MR image as well as the corresponding mask image. The discrimination term $D(img)$ in Eq. (2) and (3) should be replaced as $D(img, S(img))$.

Entropy Map. Secondly, existing template generation methods usually produce rough images and refine them after many iteration steps. To encourage the network to create more explicit images in fewer iterations, we utilize the mid-product of the segmentation task to decrease pixel/voxel-wise uncertainty. At the last layer of the segmentor S, we apply the softmax activation, which will generate the probability distribution that sums to one. That is, for each image element i, we have the label probability distribution $[P_i^1, P_i^2, ..., P_i^{N_s}]$, and we can calculate the entropy for this element. The entropy reflects how confident the segmentor is when classifying this element, which also indicates the degree of ambiguity. We use the average entropy as one of the loss terms:

$$\mathcal{L}_{entropy} = -\frac{1}{N_p} \sum_{i}^{N_p} \sum_{k}^{N_s} P_i^k \log(P_i^k), \tag{5}$$

where P_i^k is the probability for k_{th} label at i_{th} pixel. N_s is the number of segmentation labels and N_p is the overall area/volume of the image.

To conclude, the objective loss function for the generator comprises of the adversarial term, the similarity term, the magnitude term and the entropy term:

$$\mathcal{L}_G = \lambda_0 \mathcal{L}_{adv} + \lambda_1 \mathcal{L}_{sim} + \lambda_2 \mathcal{L}_{mag} + \lambda_3 \mathcal{L}_{entropy}. \tag{6}$$

3 Architecture

The framework is implemented by Pytorch [20] with Adam optimizer. The generator G is a U-Net [21], which can be divided into the encoder and the decoder. The encoder contains four consecutive downsampling blocks, each consists of a convolution layer, ReLU activation, and batch normalization (BN). The decoder contains four upsampling blocks that consist of a convolution layer, ReLU, BN, and an upsampling layer with scale = 2. Skip connections between mirror layers of the same size are applied to propagate spatial information and stabilize the training process. The discriminator D contains four downsampling convolution layers followed by BN and LeakyReLU. Then the tensor is flattened and linearly transformed to one value, which will be activated by Sigmoid. The step ratio G:D is adjusted dynamically according to the score of the discriminator. We explore the hyper-parameters and the range $\lambda_0 \in [0.05, 0.1], \lambda_1 = 10, \lambda_2 \in [0.5, 1], \lambda_3 \in [0.1, 0.2]$ works best. Lower λ_2 will result in template bias, higher λ_3 will increase the sharpness but affect the overall performance. We train the 2D model for 2×10^4 with learning rate (lr) = 1e−4 and 3D for 4×10^4 iterations.

The auxiliary segmentation network and the registration network are trained separately. The segmentation network S has four downsampling blocks, upsampling blocks, and skip connections. We train the segmentation network for 40 epochs with lr = 1e−4 using the cross-entropy loss. For the registration network, we adopt the widely-used learning-based deformable registration method VoxelMorph [3]. We train the network for 2×10^4 iterations (for 3D, 1×10^5) with negative normalization cross-correlation as the metric. The regularization parameter is set to 2.

4 Experiments

4.1 Datasets and Implementation

We use brain MRI images from OASIS [16] to conduct the template synthesis and registration experiments. The dataset consists of 425 T1 MRI scans from 416 subjects, and each scan comes with segmentation labels of 48 anatomical structures for evaluation. All scans were cropped and resampled to $176 \times 224 \times 160$ voxels, then affinely registered to one scan by simpleITK [25]. In this paper, we experiment on both extracted 2D coronal slices with the group size = 16 and 3D volumes with the group size = 4. We split the dataset into 365, 30, and 30, respectively, as training, validation, and test set for both settings. Since it would take too much time to test all groupwise combinations within the test set, we randomly generated 600 and 100 groups for 2D and 3D evaluation.

2D experiments (including training, testing stage, and baseline methods) were taken on a 6-core i7-8700K 32 GB-RAM CPU machine with NVIDIA RTX 2070 and GTX TITAN GPU. All 3D experiments were taken on an 8-core W-3225 64 GB-RAM CPU machine with 2 RTX 3090 GPU.

4.2 Evaluation Metrics

Dice Score. First, we quantify the groupwise registration results using the averaged Dice score of all pairs of subjects in the group. The score computes the overlap degree of annotated segmentations provided with the dataset: $Dice = \frac{1}{N^2} \sum_{i,j} \frac{1}{N_s} \sum_{k}^{N_s} \frac{2 \times |W_i^k \cap W_j^k|}{|W_i^k| + |W_j^k|}$, where N is the number of group size, N_s is the number of labels, W_i^k is the pixel set of k_{th} region of i_{th} warped subject. To avoid the affect of occlusion and fine particles which vary a lot between subjects, only $N_s = 19$ for 2D and 30 for 3D largest regions are applied to compute the Dice scores. Higher scores indicate better registration performance.

Smoothness. We also measure the displacement fields' smoothness to illustrate the performance of the template in the registration process. The smoothness is measured by the spatial gradients of the displacement fields: $Smooth = \frac{1}{N} \sum_{i=1}^{N} ||\nabla \phi_i||_2^2$. A smaller score refers to better performance.

Unbiasedness. To quantitatively evaluate the unbiasedness of one template, we propose to compare the standard deviation of the average absolute displacement field of each group subject when they are warped to the template image: $Bias = \sigma(\overline{|\phi_1|}, \overline{|\phi_2|}, ..., \overline{|\phi_N|})$, where ϕ_i refers to the displacement field of the i_{th} subject. A smaller bias score indicates more uniform registration among the group subjects and refers to the unbiasedness of the generated template.

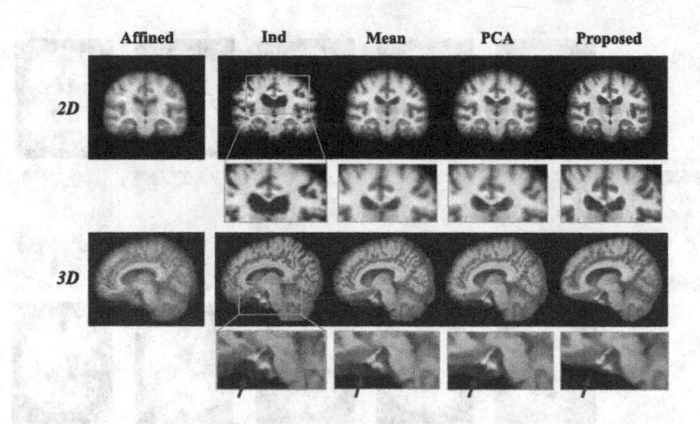

Fig. 2. Comparison of templates generated by each method for 2D and 3D MR images. For each template, there is a zoomed window. The proposed templates are clearer and sharper.

4.3 Baselines

We compare our proposed template generation framework with three baseline methods. The first template acquisition method denoted as **Ind** is to select an individual subject from the input group with the minimum intensity distance to all other subjects as the template. We also compare our method to the Group**Mean** method [15], which is the most widely used baseline method. A more recently proposed state-of-the-art method denoted as **PCA** constructs a template image using the principal components of all group subjects [8]. The registration method **VoxelMorph** [3] is set as mentioned in Sect. 3. All templates will be updated with the registration process iteratively until the Dice score declines.

4.4 Results

Performance. Figure 2 shows the 2D and 3D templates generated by each method and their zoomed patches. It can be observed that our templates are more precise at the edges. In Fig. 3, we visually displayed the GIR results of the proposed method.

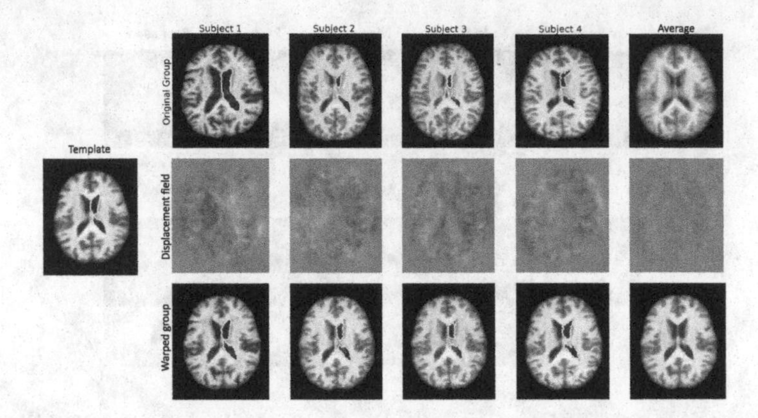

Fig. 3. Axial view of 3D MR groupwise registration results. Subjects 1–4 with the contour of lateral ventricles (in red and yellow) on the first row refer to the group subjects to be registered. The second row presents the flow fields for each subject with the value in XYZ coordinates mapped into RGB channels. The rightmost image of each row is the average. It is more colorless than the previous displacement fields because the flows in each direction cancel each other out, which means the images are warped to the center. We present the warped images and their average in the last row. (Color figure online)

We evaluate our method with three baselines Ind, Mean, and PCA using three metrics: Dice score, displacement field smoothness, and unbiasedness. The qualitative evaluation is displayed in Table 1. For the 2D experiments, our method performs the best in the aspect of Dice score (outperforms baseline methods Ind, Mean, PCA by 3.5%, 1.2% and 1.6%) and unbiasedness (outperforms by a larger margin). While our performance on displacement smoothness is slightly worse than Mean (1.257 and 1.235). Our method works even better in the 3D setting. The proposed method outperforms baseline methods in terms of all aspects. The improvement on Dice Score is at least 0.9%, Smoothness Score at least 7.7%, and Bias Score 12.8%. Furthermore, it is worth noting that for both datasets, our method has much shorter inference time than baselines that may have comparable performance to ours.

Ablation Studies. To verify each module's necessity in our template generation method, we perform the ablation study and present the results in Table 2. Trial 1 refers to the basic GAN scheme of a generator and a discriminator. This trial is quite like the individual groupwise registration method, as the generator constructs authentic images. In trial 2, we add a deformable registration network to warp group subjects and calculate their similarity. In trial 3, we add a constraint on the magnitude of the displacement field. In addition to the improved qualitative scores, experiments show this $\mathcal{L}_{mag}$ term can also keep the generated template close to the latent center space; without this term, the template would be pretty different from the group average (like being smaller or larger). Trials 4 and 5 add the segmentation mask as auxiliary information for the discriminator to discern. The proposed method trial 5 adds the entropy term and achieves the best performance.

Table 1. Performance of our method on 2D and 3D MR images compared with baseline methods in terms of average Dice score, smoothness and unbiasedness. All best scores in bold significantly outperform the second best (p-value $\ll 0.05$ in a paired t-test).

Methods		Dice Score %	Smoothness %	Bias Score %	Time (sec)	
					CPU	GPU
2D	Affine	55.53 ± 3.269	–	–	–	–
2D	Ind	65.98 ± 2.022	1.415 ± 0.109	16.50 ± 5.212	0.964	0.619
2D	Mean	67.50 ± 1.500	$\mathbf{1.235 \pm 0.056}$	14.96 ± 3.840	5.629	4.512
2D	PCA	67.24 ± 1.812	1.320 ± 0.084	12.73 ± 2.117	4.579	4.422
2D	**Proposed**	$\mathbf{68.31 \pm 1.538}$	1.257 ± 0.060	$\mathbf{7.184 \pm 2.188}$	0.829	0.603
3D	Affine	52.11 ± 6.205	–	–	–	–
3D	Ind	65.10 ± 4.937	1.383 ± 0.227	17.936 ± 3.780	119.6	6.246
3D	Mean	66.79 ± 4.415	1.266 ± 0.164	6.401 ± 2.835	291.8	35.22
3D	PCA	65.95 ± 4.774	1.295 ± 0.186	6.141 ± 2.152	163.0	19.14
3D	**Proposed**	$\mathbf{67.39 \pm 4.270}$	$\mathbf{1.169 \pm 0.098}$	$\mathbf{5.352 \pm 2.528}$	123.9	13.16

Table 2. Ablation studies on the 2D validation dataset. Trial 1–3 use basic GAN structure without segmentation information. Trial 5 is the final method.

Trial	Methods	Dice %	Smooth %	Bias %
0	Affine	55.17	–	–
1	Basic GAN	65.52	2.24	11.7
2	Basic GAN + $\mathcal{L}_{sim}$	67.78	1.75	8.44
3	Basic GAN + $\mathcal{L}_{sim}$ + $\mathcal{L}_{mag}$	67.97	1.61	7.48
4	Seg-assisted GAN + $\mathcal{L}_{sim}$ + $\mathcal{L}_{mag}$	68.36	1.47	5.81
5	Seg-assisted GAN + $\mathcal{L}_{sim}$ + $\mathcal{L}_{mag}$ + $\mathcal{L}_{entropy}$	68.65	1.25	5.92

5 Conclusion

In this paper, we have presented a learning-based unbiased template synthesis method for groupwise medical image registration. The method utilizes the generative adversarial network scheme with the auxiliary registration network and the segmentor to synthesize a realistic and unbiased template. The new method can improve the registration results, and the template can effectively reflect the features of the given group of medical images. We conducts sets of experiments on 2D/3D MR images and evaluate the results from accuracy, smoothness, and unbiasedness. The results demonstrate that our method can achieve an overall state-of-the-art performance.

Meanwhile, there are still some issues to be solved in the future. First, the training of GAN is complex and unstable. In our implementation, the training ratio of the generator and discriminator is adjusted automatically according to the discriminator's scores during training. In the future, we will explore different dynamic adaptation strategies for the GAN scheme. Second, the group size is fixed for each trained model, making it not easy to generalize to groups of arbitrary sizes. We believe this can be solved by introducing some hierarchical processing methods that divide a large group into smaller subgroups and then perform GIR. Furthermore, we will explore the latent space representation to perform groupwise image processing flexibly.

References

1. Agier, R., Valette, S., Kéchichian, R., Fanton, L., Prost, R.: Hubless keypoint-based 3D deformable groupwise registration. Med. Image Anal. **59**, 101564 (2020)
2. Auzias, G., et al.: Diffeomorphic brain registration under exhaustive sulcal constraints. IEEE Trans. Med. Imaging **30**(6), 1214–1227 (2011)
3. Balakrishnan, G., Zhao, A., Sabuncu, M.R., Guttag, J., Dalca, A.V.: Voxelmorph: a learning framework for deformable medical image registration. IEEE Trans. Med. Imaging **38**(8), 1788–1800 (2019)
4. Ben-Cohen, A., Klang, E., Raskin, S.P., Amitai, M.M., Greenspan, H.: Virtual PET images from CT data using deep convolutional networks: initial results. In: Tsaftaris, S.A., Gooya, A., Frangi, A.F., Prince, J.L. (eds.) SASHIMI 2017. LNCS, vol. 10557, pp. 49–57. Springer, Cham (2017). https://doi.org/10.1007/978-3-319-68127-6_6
5. Bi, L., Kim, J., Kumar, A., Feng, D., Fulham, M.: Synthesis of positron emission tomography (PET) images via multi-channel generative adversarial networks (GANs). In: Cardoso, M.J., et al. (eds.) CMMI/SWITCH/RAMBO -2017. LNCS, vol. 10555, pp. 43–51. Springer, Cham (2017). https://doi.org/10.1007/978-3-319-67564-0_5
6. Cai, J., Zhang, Z., Cui, L., Zheng, Y., Yang, L.: Towards cross-modal organ translation and segmentation: a cycle-and shape-consistent generative adversarial network. Med. Image Anal. **52**, 174–184 (2019)
7. Chartsias, A., Joyce, T., Dharmakumar, R., Tsaftaris, S.A.: Adversarial image synthesis for unpaired multi-modal cardiac data. In: Tsaftaris, S.A., Gooya, A., Frangi, A.F., Prince, J.L. (eds.) SASHIMI 2017. LNCS, vol. 10557, pp. 3–13. Springer, Cham (2017). https://doi.org/10.1007/978-3-319-68127-6_1
8. Che, T., et al.: DGR-Net: deep groupwise registration of multispectral images. In: Chung, A.C.S., Gee, J.C., Yushkevich, P.A., Bao, S. (eds.) IPMI 2019. LNCS, vol. 11492, pp. 706–717. Springer, Cham (2019). https://doi.org/10.1007/978-3-030-20351-1_55
9. Dalca, A.V., Rakic, M., Guttag, J., Sabuncu, M.R.: Learning conditional deformable templates with convolutional networks. arXiv preprint arXiv:1908.02738 (2019)
10. Dong, P., Cao, X., Yap, P.T., Shen, D.: Fast groupwise registration using multilevel and multi-resolution graph shrinkage. Sci. Rep. **9**(1), 1–12 (2019)

11. Fonov, V., et al.: Unbiased average age-appropriate atlases for pediatric studies. Neuroimage **54**(1), 313–327 (2011)
12. Goodfellow, I., et al.: Generative adversarial nets. In: Advances in Neural Information Processing Systems, pp. 2672–2680 (2014)
13. He, Z., Chung, A.C.: Unsupervised end-to-end groupwise registration framework without generating templates. In: 2020 IEEE International Conference on Image Processing (ICIP), pp. 375–379. IEEE (2020)
14. Jaderberg, M., Simonyan, K., Zisserman, A., Kavukcuoglu, K.: Spatial transformer networks. arXiv preprint arXiv:1506.02025 (2015)
15. Joshi, S., Davis, B., Jomier, M., Gerig, G.: Unbiased diffeomorphic atlas construction for computational anatomy. Neuroimage **23**, S151–S160 (2004)
16. Marcus, D.S., Wang, T.H., Parker, J., Csernansky, J.G., Morris, J.C., Buckner, R.L.: Open access series of imaging studies (OASIS): cross-sectional MRI data in young, middle aged, nondemented, and demented older adults. J. Cogn. Neurosci. **19**(9), 1498–1507 (2007)
17. Mok, T.C.W., Chung, A.C.S.: Large deformation diffeomorphic image registration with Laplacian pyramid networks. In: Martel, A.L., et al. (eds.) MICCAI 2020. LNCS, vol. 12263, pp. 211–221. Springer, Cham (2020). https://doi.org/10.1007/978-3-030-59716-0_21
18. Nie, D., Cao, X., Gao, Y., Wang, L., Shen, D.: Estimating CT image from MRI data using 3D fully convolutional networks. In: Carneiro, G., et al. (eds.) LABELS/DLMIA -2016. LNCS, vol. 10008, pp. 170–178. Springer, Cham (2016). https://doi.org/10.1007/978-3-319-46976-8_18
19. O'Donnell, L.J., Wells, W.M., Golby, A.J., Westin, C.-F.: Unbiased groupwise registration of white matter tractography. In: Ayache, N., Delingette, H., Golland, P., Mori, K. (eds.) MICCAI 2012. LNCS, vol. 7512, pp. 123–130. Springer, Heidelberg (2012). https://doi.org/10.1007/978-3-642-33454-2_16
20. Paszke, A., et al.: Pytorch: an imperative style, high-performance deep learning library. In: Advances in Neural Information Processing Systems, vol. 32, pp. 8024–8035. Curran Associates, Inc. (2019)
21. Ronneberger, O., Fischer, P., Brox, T.: U-Net: convolutional networks for biomedical image segmentation. In: Navab, N., Hornegger, J., Wells, W.M., Frangi, A.F. (eds.) MICCAI 2015. LNCS, vol. 9351, pp. 234–241. Springer, Cham (2015). https://doi.org/10.1007/978-3-319-24574-4_28
22. Sabuncu, M.R., Balci, S.K., Shenton, M.E., Golland, P.: Image-driven population analysis through mixture modeling. IEEE Trans. Med. Imaging **28**(9), 1473–1487 (2009)
23. Wang, Y., et al.: Semisupervised tripled dictionary learning for standard-dose pet image prediction using low-dose pet and multimodal MRI. IEEE Trans. Biomed. Eng. **64**(3), 569–579 (2016)
24. Wu, G., Jia, H., Wang, Q., Shen, D.: Sharpmean: groupwise registration guided by sharp mean image and tree-based registration. Neuroimage **56**(4), 1968–1981 (2011)
25. Yaniv, Z., Lowekamp, B.C., Johnson, H.J., Beare, R.: Simpleitk image-analysis notebooks: a collaborative environment for education and reproducible research. J. Digit. Imaging **31**(3), 290–303 (2018). https://doi.org/10.1007/s10278-017-0037-8
26. Yu, B., Wang, Y., Wang, L., Shen, D., Zhou, L.: Medical image synthesis via deep learning. In: Lee, G., Fujita, H. (eds.) Deep Learning in Medical Image Analysis. AEMB, vol. 1213, pp. 23–44. Springer, Cham (2020). https://doi.org/10.1007/978-3-030-33128-3_2

27. Yu, B., Zhou, L., Wang, L., Shi, Y., Fripp, J., Bourgeat, P.: Ea-GANs: edge-aware generative adversarial networks for cross-modality MR image synthesis. IEEE Trans. Med. Imaging **38**(7), 1750–1762 (2019)
28. Zhang, T., et al.: SkrGAN: sketching-rendering unconditional generative adversarial networks for medical image synthesis. In: Shen, D., et al. (eds.) MICCAI 2019. LNCS, vol. 11767, pp. 777–785. Springer, Cham (2019). https://doi.org/10.1007/978-3-030-32251-9_85
29. Zhang, Z., Yang, L., Zheng, Y.: Translating and segmenting multimodal medical volumes with cycle-and shape-consistency generative adversarial network. In: Proceedings of the IEEE Conference on Computer Vision and Pattern Recognition, pp. 9242–9251 (2018)

The Role of MRI Physics in Brain Segmentation CNNs: Achieving Acquisition Invariance and Instructive Uncertainties

Pedro Borges[1,2(✉)], Richard Shaw[1,2], Thomas Varsavsky[1,2], Kerstin Klaser[2], David Thomas[3], Ivana Drobnjak[1], Sebastien Ourselin[2], and M. Jorge Cardoso[2]

[1] Department of Medical Physics and Biomedical Engineering, UCL, London, UK
p.borges.17@ucl.ac.uk

[2] School of Biomedical Engineering and Imaging Sciences, KCL, London, UK

[3] Dementia Research Centre, UCL, London, UK

Abstract. Being able to adequately process and combine data arising from different sites is crucial in neuroimaging, but is difficult, owing to site, sequence and acquisition-parameter dependent biases. It is important therefore to design algorithms that are not only robust to images of differing contrasts, but also be able to generalise well to unseen ones, with a quantifiable measure of uncertainty. In this paper we demonstrate the efficacy of a physics-informed, uncertainty-aware, segmentation network that employs augmentation-time MR simulations and homogeneous batch feature stratification to achieve acquisition invariance. We show that the proposed approach also accurately extrapolates to out-of-distribution sequence samples, providing well calibrated volumetric bounds on these. We demonstrate a significant improvement in terms of coefficients of variation, backed by uncertainty based volumetric validation.

1 Introduction

Magnetic Resonance Imaging (MRI) is one of the most widespread neuroimaging techniques owing to its excellent soft tissue contrast, boasting great versatility in highlighting different regions and pathologies by means of sequence selection. As a consequence, a significant body of work has emerged developing accurate processing algorithms for MR images that may arise from different sites and acquisition sequence parameters. There are those works that focus on achieving algorithms that can generalise well to all contrasts. Traditional and largely widespread techniques include probabilistic generative models [2] and multi-atlas fusion methods [17]. However, the former has strong assumptions on label intensity distributions, and the latter is predicated on lengthy processing times due to its dependence on image registration. Recent works using convolutional neural networks (CNNs), such as Billot *et al.* [3], tackle contrast agnosticism by employing a Bayesian generative segmentation model that synthesises images

D. Svoboda et al. (Eds.): SASHIMI 2021, LNCS 12965, pp. 67–76, 2021.
https://doi.org/10.1007/978-3-030-87592-3_7

containing multiple different contrasts. Jog *et al.* [12] devise an approach by which networks can be made to generalise to unseen contrasts by predicting pulse sequence parameters from such images, and simulating images of that contrast by using labelled multiparametric map datasets. Pham *et al.* [16] employ an iterative approach involving a dual segmentation-synthesis model, whereby images of unseen contrasts are segmented, used to train a synthesis network that in turn is used to generate new images of the unseen contrast from the labels in the original training set. It is important to note that, while these methods are able to segment data from unseen sites with some degree of accuracy, they do not model the interaction between acquisition parameters and the underlying anatomy explicitly - they segment what they see and not the true anatomy.

This leads to those methods that seek to harmonise measurements across sites by directly accounting for such covariates as scanner and site bias, and sequence contrast variabilities, e.g. ComBat [13] is a Bayesian framework designed to account for experimental variabilities that has been applied to cortical thickness harmonisation [7]. These classes of techniques, however, operate directly on extracted volumetric measurements and not on the images. Harmonisation has also been tackled with CycleGANs [19,20] and domain adaptation approaches [5].

Recent work [4] proposed a means to directly introduce the physics of the MR acquisition process directly into deep learning networks in combination with pre-generated synthetic MR images based on multi-parametric MR maps (MPMs). This work achieves some agnosticism to the underlying physics by demonstrating that generated segmentations are more consistent volumetrically. This method, however, does not enforce volumetric consistency across contrasts, and has not been show to extrapolate to out of distribution sequence parameters.

Changes in MRI acquisition parameters alter the tissue contrast, thus impacting the algorithmic ability to accurately segment images; this can be modeled via uncertainty estimation. Here, we propose to model both epistemic (ability of the model to know) and aleatoric (unknowns of the data) uncertainties. Building on existing work [4], we also introduce a new training approach and consistency loss across realisations of MRI contrasts, allowing the model to appropriately disentangle the anatomical phenotype and the MRI physics, and extrapolate to unseen contrasts without sacrificing segmentation quality.

2 Methods

Borges *et al.* [4] proposed that a network could be made resilient to changes in the physics parameters, and therefore be able to appropriately segment data produced by different sequences. This was achieved by generating simulated data, and passing this imaging data and associated MRI parameters to a CNN. In order to train against a "Physics Gold Standard" (PGS), i.e. a true model of the anatomy that is not influenced by the choice of acquisition parameters, the authors used a Gaussian Mixture Model of literature sourced tissue parameters for grey matter (GM), white matter (WM), and cerebrospinal fluid (CSF) on their quantitative MPMs. We build on this work and improve the algorithmic

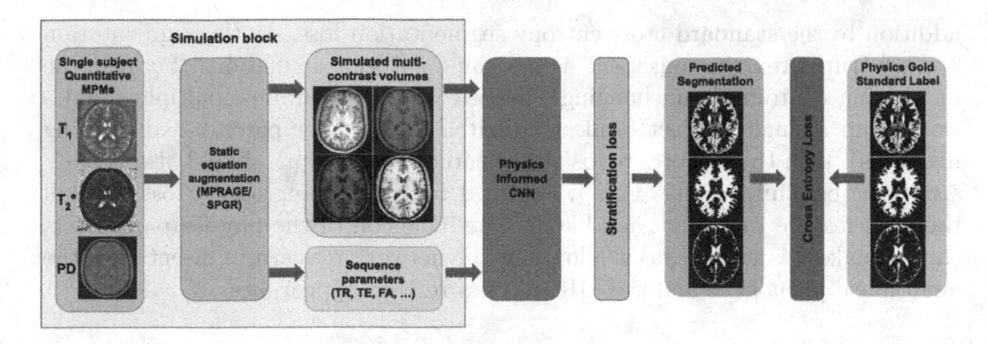

Fig. 1. The training pipeline with proposed new additions of single subject batch stratification and accompanying L_2 feature maps loss, and training time image simulation.

robustness, ameliorating image quality, segmentation volume consistency, and validating within and out of distribution samples paired with uncertainty derived errors.

2.1 Network Architecture

In Borges *et al.* [4], the injection of the physics parameters into the network is done via the inclusion of two fully connected layers whose output is tiled and concatenated to the ante-penultimate convolutional layer output. We adopt a similar strategy, but take the added step of also tiling this output to an earlier region of the network, immediately preceding the first down-sampling layer. We argue that knowledge of the physics is potentially valuable information in the encoding portion of the network, and that this allows it to better disentangle the physics parameters and the subject's phenotype.

We moved to adopt the nn-UNet architecture [10]. All networks were trained with batch size 4, on 3D patches of size 128^3 sampled from the simulated volumes. Networks were trained with a learning rate of 10^{-4} until convergence, where convergence is defined as 7 epochs elapsing without an improvement in validation metrics, Dice score combined with coefficient of variation (CoV). We made use of two main frameworks for this work, TorchIO [15], and MONAI [1].

As the proposed method requires multi-parametric data to train the model, a more scarce resource in large numbers, a dataset comprised of 18 subjects were used for training, four for validation, and five for inference/testing.

2.2 Stratification and Batch Homogeneity

We seek to further enforce volumetric consistencies vis-à-vis same-subject realisations generated using different sequence parameters. We therefore propose a batch stratification approach where each batch contains multiple realisations of images from a single subject. This allows for the addition of a stratification loss over the batch features of the penultimate layer of our network, which acts in addition to the standard cross-entropy segmentation loss. As the segmentation ground truths remain consistent across same subject simulations (because the underlying anatomy is unchanging), if a given batch contains multiple simulations from a single subject (and same patch location for patch-based training networks), then the features maps at the end of the network should also be consistent across simulations. This is enforced by introducing an L_2 loss over all the final feature maps for each batch. The inclusion of the physics parameters should make this tenable, as it allows for the network to learn to disentangle the anatomical phenotype and the MRI-physics related appearance.

2.3 Casting Simulation as an Augmentation Layer

We adopt the same static equation multi-parametric map based simulation approach as Jog *et al.* [11], focusing on MPRAGE and SPGR sequences. The SPGR equation describing the signal b_S per voxel, x is:

$$b_S(x) = G_S PD(x) sin\theta \frac{1 - e^{-\frac{TR}{T_1(x)}}}{1 - \cos\theta e^{-\frac{TR}{T_1(x)}}} e^{-\frac{TE}{T_2^*(x)}}, \tag{1}$$

where G_S is the scanner gain, TR the repetition time, T_1 the longitudinal relaxation time, TE the echo time, and T_2^* the transverse relaxation time.

Similarly, for MPRAGE:

$$b_M(x) = G_S PD(x) \left(1 - \frac{2e^{\frac{-TI}{T_1(x)}}}{1 + e^{\frac{-(TI+TD+\tau)}{T_1(x)}}}\right), \tag{2}$$

where TD the delay time, and τ the slice imaging time.

Unlike in [4], where simulated volumes are all generated prior to network training, we implemented the static equation simulation layer as an augmentation layer. Such a layer takes as input a 4D multi-parametric map, a protocol type, and a range of relevant parameters to randomly sample from, producing N (batch size) simulated volumes. This layer-based batch approach is compatible with our posed stratification model, as all generated volumes per batch belong to the same subject, permitting the utilisation of the within-batch feature consistency loss. The full training pipeline is depicted in Fig. 1.

2.4 Uncertainty Modelling

We opt to incorporate uncertainty modelling in our framework to obtain volumetric bounds on our segmentations. We model the aleatoric uncertainty via explicit loss attenuation [14]. We modify our network architecture to include an additional convolutional block that branches off the final upsampling layer. This branch models the aleatoric uncertainty, $\sigma_{Het}^W(x)$. This modifies the cross-entropy loss function accordingly:

$$\hat{x}_{i,t} = f_i^W + \epsilon_t, \qquad \epsilon_t \sim \mathcal{N}(0, (\sigma_i^W)^2) \tag{3}$$

$$\mathcal{L} = \sum_i log \frac{1}{T} \sum_t c_i(-\hat{x}_{i,t,c} + log \sum_{c'} e^{\hat{x}_{i,t,c'}}) \tag{4}$$

Where $\hat{x}_{i,t}$ are the task logits (f_i^W) summed with a noise sample of standard deviation equal to the predicted σ_i^W per voxel; T denotes the number of stochastic passes per input, and σ^W is defined for every voxel, per class, c. This allows for the easy extraction of volumetric bounds by repeatedly sampling from additive logit noise distributions to produce new segmentations.

The epistemic uncertainty is modelled using test-time Monte Carlo sampling via dropout. Dropout is commonly used as a regularisation technique [18], but also allows for the approximate Bayesian posterior sampling of segmentations by maintaining the random neuron switching at test-time [8]. We set a dropout level of 0.5 in all layers except for the input layer, where it is set to 0.05.

3 Experiments

3.1 Data

We make use of a 27 subject multi-parametric early onset Alzheimer dataset, the same as in [4], for the purpose of simulating images which are used for training, validating, and testing of our models, all of which are registered to MNI space rigidly. The images contain maps of the longitudinal and effective transverse magnetisation relaxation, R_1 and R_2^*, proton density, PD, and magnetisation transfer, MT. The details concerning quantitative map creation can be found in [9]. The static equation models we employ feature T_1 (inverse of R_1), T_2^* (inverse of R_2^*), and PD.

3.2 Simulation Sequence Details

To allow for direct comparability, we limited the ranges of the relevant parameters for simulated images at training time to those stipulated in the original work, i.e. inversion time (TI) = [600–1200] ms for MPRAGE, repetition time (TR) = [15–100] ms, echo time (TE) = [4–10] ms, and flip angle (FA) = [15–75] degrees for SPGR. For each subject, a single "Physics Gold Standard" (PGS) segmentation was used across the associated synthesized images, generated using the same process and literature values as in the original work [4].

4 Annealing Study: Robustness and Quality Analysis

To ascertain the contributions of the two main additions to the underlying method, we carry out an annealing study, whereby we analyse the incremental performance increases in terms of volume consistency and Dice score, with the addition of each change. We begin with a complete physics-agnostic baseline, i.e. a standard 3D nn-UNet trained with pre-generated data (Baseline), followed by the original physics method (Phys-Base), followed by Phys-Base with the addition of batch stratification (Phys-Strat), followed lastly by Phys-Strat with the addition of the simulation augmentation scheme (Phys-Strat-Aug).

Table 1. Mean dice scores for Baseline, Phys-Base, Phys-Strat, and Phys-Strat-Aug on segmentation task, across inference subjects. All dice scores are estimated against a Physics Gold Standard. Standard deviations quoted in brackets. Bold values represent statistically best performances.

Experiments	Sequence dice scores							
	MPRAGE				SPGR			
	GM		WM		GM		WM	
	IoD	OoD	IoD	OoD	IoD	OoD	IoD	OoD
Baseline	0.966 (0.005)	0.956 (0.006)	0.953 (0.002)	0.934 (0.002)	0.878 (0.021)	0.872 (0.008)	0.893 (0.023)	0.873 (0.011)
Phys-Base	**0.971 (0.007)**	0.964 (0.009)	**0.964 (0.008)**	**0.959 (0.011)**	0.911 (0.020)	0.872 (0.050)	0.912 (0.021)	0.880 (0.092)
Phys-Strat	**0.970 (0.005)**	**0.969 (0.005)**	0.958 (0.004)	**0.957 (0.005)**	**0.929 (0.015)**	**0.911 (0.011)**	**0.922 (0.021)**	0.894 (0.040)
Phys-Strat-Aug	**0.971 (0.004)**	**0.971 (0.005)**	**0.962 (0.003)**	**0.960 (0.004)**	**0.930 (0.016)**	**0.913 (0.019)**	**0.921 (0.015)**	**0.899 (0.019)**

Table 2. Coefficients of variation (CoV) for Baseline, Phys-Base, Phys-Strat, and Phys-Strat-Aug on segmentation task, averaged across test subjects. Standard deviations quoted in brackets. Bold values represent statistically best performances.

Experiments	Sequence CoVs (x10^3)							
	MPRAGE				SPGR			
	GM		WM		GM		WM	
	IoD	OoD	IoD	OoD	IoD	OoD	IoD	OoD
Baseline	6.39 (0.87)	22.50 (4.08)	14.94 (1.71)	51.12 (7.11)	61.91 (7.61)	170.10 (31.32)	32.57 (11.98)	158.93 (16.83)
Phys-Base	2.72 (2.12)	14.67 (7.30)	3.28 (2.01)	28.10 (3.98)	77.22 (34.44)	127.22 (18.61)	20.77 (9.35)	264.80 (8.52)
Phys-Strat	0.71 (0.23)	6.15 (1.51)	0.53 (0.25)	**3.67 (1.34)**	21.83 (0.83)	59.78 (13.31)	8.60 (0.64)	59.19 (11.25)
Phys-Strat-Aug	**0.42 (0.22)**	**4.74 (1.30)**	**0.51 (0.23)**	**3.65 (0.62)**	**15.76 (1.18)**	**28.88 (9.74)**	**7.12 (0.45)**	**44.78 (4.22)**

We extend our volumetric consistency analysis by analysing out of distribution (OoD) samples. In this instance they are defined as simulated images whose sequence parameters lie outside of the training range. This not only results in images of unfamiliar contrasts, but also unseen parameters that are fed into the physics branch of the network. If our method has truly attained a measure of sequence invariance then it should be expected that both segmentation quality and volume consistency are maintained as the network should be able to extrapolate from the provided values. For MPRAGE, the OoD range is extended to [100–2000] ms, while for SPGR, the TR is extended to [10–200] ms, TE is extended to [2–20] ms, and FA is extended to [5–90] degrees.

Table 1 and Table 2 show Dice and CoV performances, respectively. We carry out signed-rank Wilcoxon tests to test for statistically significant improvements, and bold the best model (p-value < 0.01). Tests are carried out on CoV and Dice scores independently of each other. In instances where models may outperform baselines but are not statistically significantly different from each other, we bold both. We verify an incremental gain in CoV and Dice with each added feature, the most pronounced of which results from the addition of the stratification loss, in terms of both in and out of distribution CoVs. This is expected, as directly optimising for consistency across realisations of the same subject should more strongly enforce volume consistency, enhancing the physics invariance.

Phys-Strat-Aug boasts the best performance overall, significantly outperforming both Baseline and Phys-Base with regards to CoV. Compared to Phys-Strat, the differences are not always statistically better for MPRAGE, but are so for SPGR. With more parameters at play, an augmentation scheme should become more relevant, as sampling from the parameter space should lead to a greater extrapolating ability, as the network is no longer constrained to learn from a more discrete training set, and will experience more varied realisations.

Figure 2 shows some qualitative results, in and out of distribution segmentation comparisons between Baseline and Phys-Strat-Aug, to convey the consistency the latter is able to achieve without compromising segmentation quality.

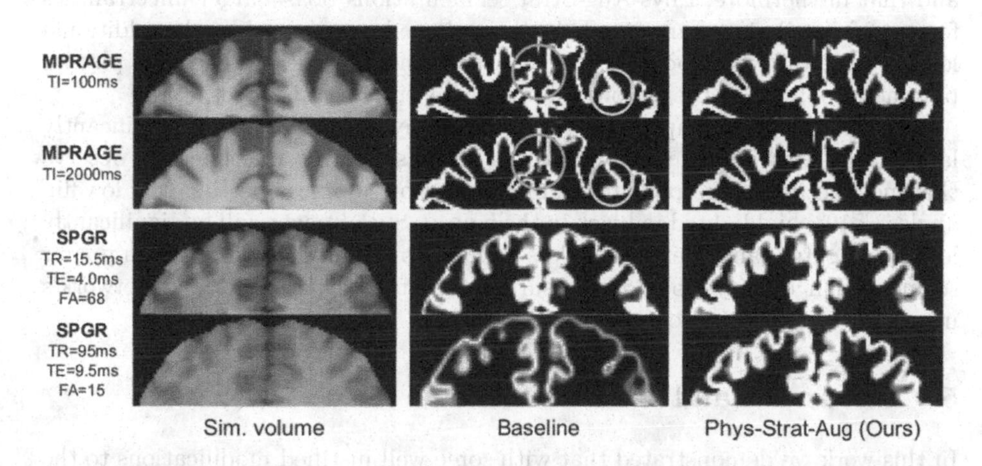

Fig. 2. Baseline and Phys-Strat-Aug comparisons. Comparing out-of-distribution MPRAGE (Top two rows) and SPGR (Bottom two rows) GM segmentations from the proposed and baseline methods. Blue circles highlight examples of significant gyrus variability. Orange circles denote regions of segmentation differences between protocols. (Color figure online)

4.1 Uncertainty Measures and Volumetric Bounds

Given Phys-Strat-Aug's superior performance, we train only two epistemic uncertainty models, and two aleatoric uncertainty models with this pipeline, one of each for this pipeline and a complete baseline.

At test-time we extract 50 aleatoric volume samples, and 50 epistemic volume samples for each of the networks, for both in and out of distribution simulated images. We verify that the aleatoric samples do not contribute significantly to the volume variance in comparison to its epistemic counterpart, (an observation that was also been verified in [6]) and therefore omit it in our volumetric analysis.

Figure 3 showcases white matter volume variations for MPRAGE and SPGR sequences, for the extended out of distribution parameter ranges, for Baseline and Phys-Strat-Aug experiments, for a single subject. For the SPGR plot, we order the points based on volumetric consistency for each experiment, thus highlighting outliers. In both instances we observe a much greater consistency in volume for Phys-Strat-Aug, itself a reflection of the aforementioned CoV results. Using the calibrated volumetric method described in [6] allows us to calculate volume percentiles for each set of dropout samples, and the errors represent the volumetric interquartile range (IQR).

The errors for the baseline do not vary in any statistically significant manner, for either sequence or tissue, independent of any volume deviation. It is a different matter for Phys-Strat-Aug, however. Specifically, for MPRAGE, we note that uncertainties are consistently larger for Phys-Strat-Aug compared to baseline, and that furthermore, Phys-Aug-Strat segmentations boast larger uncertainties for out of distribution samples. This can perhaps be explained by the additional level of uncertainty introduced by the physics, and how the presence of a physics parameter outside of the "known" further exacerbates this effect.

For SPGR, all the apparent outliers for Phys-Strat-Aug have significantly larger associated errors, while this is not the case for the Baseline. We observe that most outliers correspond to out of distribution samples boasting very low flip angles ($<10°$, highlighted in black in the figure). Such images will be significantly less T_1-weighted, and therefore be less familiar to the models, resulting in poorer segmentation quality, so it is reassuring that the physics-informed network's uncertainty around these samples is larger.

5 Discussion and Conclusions

In this work we demonstrated that with some well justified modifications to the training pipeline, a physics-informed network can achieve extremely constrained tissue segmentations across a wide range of contrasts, across all tissue types and investigated sequences; thus strengthening its harmonisation capabilities.

Furthermore, we also showed that it can suitably generalise to unseen domains, while maintaining volume consistency without compromising segmentation quality, and is validated by accurately quantifying the volumetric uncertainty. The uncertainty estimates further suggest that the physics knowledge grants the model an additional level of safety, as volumetric uncertainties proved to be larger for out of distribution parameter generated images.

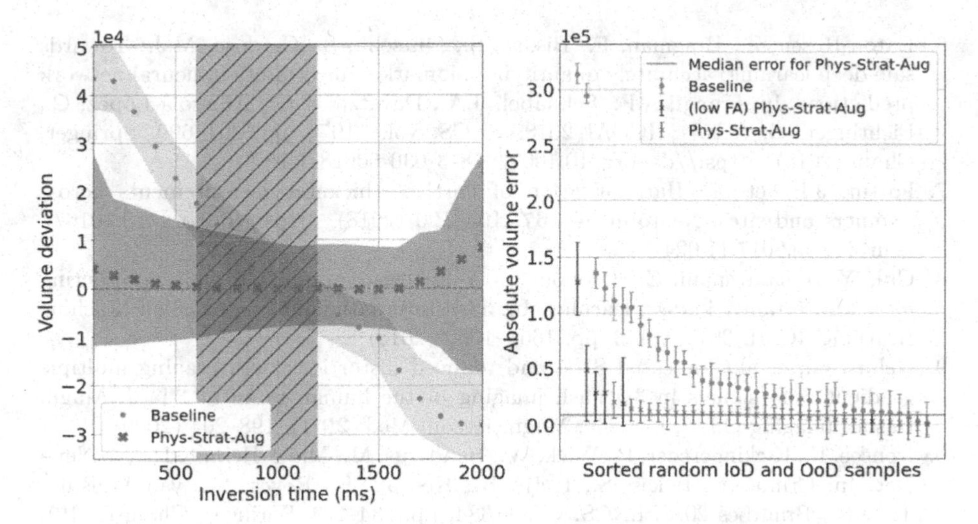

Fig. 3. Volume consistency for WM for complete baseline and Phys-Strat-Aug, for example subject. Filled plots/ Error bars correspond to IQR volumes. Left: MPRAGE. The dashed grey region denotes the TI training time parameter range (600–1200 ms). Right: SPGR. Black points denote samples with FA lower than 10° for Phys-Strat-Aug.

The method is admittedly limited by those sequences that can be aptly represented as a static equation, but we argue that at the very least, for the purposes of contrast agnosticism, a wide enough range of realistic contrasts can be generated with currently implemented sequences, which should allow for our method to generalise further. Future work will therefore involve testing of our method on multiple external datasets to ascertain generalisability and the exploration of techniques that may allow for the modelling of MR artifacts such as movement and B_0 inhomogeneities, to enhance our model's utility.

Acknowledgements. This project was funded by the Wellcome Flagship Programme (WT213038/Z/18/Z) and Wellcome EPSRC CME (WT203148/Z/16/Z).

References

1. Project MONAI. https://doi.org/10.5281/zenodo.4323059
2. Ashburner, J., Friston, K.J.: Unified segmentation. Neuroimage **26**(3), 839–851 (2005). https://doi.org/10.1016/j.neuroimage.2005.02.018
3. Billot, B., Greve, D., Van Leemput, K., Fischl, B., Iglesias, J.E., Dalca, A.V.: A learning strategy for contrast-agnostic MRI segmentation. arXiv (2020)
4. Borges, P., et al.: Physics-informed brain MRI segmentation. In: Burgos, N., Gooya, A., Svoboda, D. (eds.) SASHIMI 2019. LNCS, vol. 11827, pp. 100–109. Springer, Cham (2019). https://doi.org/10.1007/978-3-030-32778-1_11
5. Papież, B.W., Namburete, A.I.L., Yaqub, M., Noble, J.A. (eds.): MIUA 2020. CCIS, vol. 1248. Springer, Cham (2020). https://doi.org/10.1007/978-3-030-52791-4

6. Eaton-Rosen, Z., Bragman, F., Bisdas, S., Ourselin, S., Cardoso, M.J.: Towards safe deep learning: accurately quantifying biomarker uncertainty in neural network predictions. In: Frangi, A.F., Schnabel, J.A., Davatzikos, C., Alberola-López, C., Fichtinger, G. (eds.) MICCAI 2018. LNCS, vol. 11070, pp. 691–699. Springer, Cham (2018). https://doi.org/10.1007/978-3-030-00928-1_78
7. Fortin, J.P., et al.: Harmonization of cortical thickness measurements across scanners and sites. NeuroImage **167**, 104–120 (2018). https://doi.org/10.1016/j.neuroimage.2017.11.024
8. Gal, Y., Ghahramani, Z.: Dropout as a Bayesian approximation: representing model uncertainty in deep learning. In: 33rd International Conference on Machine Learning, ICML 2016, vol. 3, pp. 1651–1660 (2015)
9. Helms, G., et al.: Increased SNR and reduced distortions by averaging multiple gradient echo signals in 3D flash imaging of the human brain at 3T. J. Magn. Reson. Imaging Off. J. Int. Soc. Magn. Reson. Med. **29**(1), 198–204 (2009)
10. Isensee, F., Kickingereder, P., Wick, W., Bendszus, M., Maier-Hein, K.H.: No New-Net. In: Crimi, A., Bakas, S., Kuijf, H., Keyvan, F., Reyes, M., van Walsum, T. (eds.) BrainLes 2018. LNCS, vol. 11384, pp. 234–244. Springer, Cham (2019). https://doi.org/10.1007/978-3-030-11726-9_21
11. Jog, A., Carass, A., Roy, S., Pham, D.L., Prince, B., Amod, J.L.: MR image synthesis by contrast learning on neighborhood ensembles. Med. Image Anal. **24**, 63–76 (2015). https://doi.org/10.1016/j.media.2015.05.002
12. Jog, A., Hoopes, A., Greve, D.N., Van Leemput, K., Fischl, B.: PSACNN: pulse sequence adaptive fast whole brain segmentation. NeuroImage **199**, 553–569 (2019)
13. Johnson, W.E., Li, C., Rabinovic, A.: Adjusting batch effects in microarray expression data using empirical Bayes methods. Biostatistics **8**(1), 118–127 (2007). https://doi.org/10.1093/biostatistics/kxj037
14. Kendall, A., Gal, Y.: What uncertainties do we need in bayesian deep learning for computer vision? Technical report (2017)
15. Pérez-García, F., Sparks, R., Ourselin, S.: TorchIO: a Python library for efficient loading, preprocessing, augmentation and patch-based sampling of medical images in deep learning. Comput. Methods Prog. Biomed. **208**, 106236 (2020)
16. Pham, D.L., Chou, Y.-Y., Dewey, B.E., Reich, D.S., Butman, J.A., Roy, S.: Contrast adaptive tissue classification by alternating segmentation and synthesis. In: Burgos, N., Svoboda, D., Wolterink, J.M., Zhao, C. (eds.) SASHIMI 2020. LNCS, vol. 12417, pp. 1–10. Springer, Cham (2020). https://doi.org/10.1007/978-3-030-59520-3_1
17. Sabuncu, M.R., Yeo, B.T., Leemput, K.V., Fischl, B., Golland, P.: A generative model for image segmentation based on label fusion. IEEE Trans. Med. Imaging **29**, 1714–1729 (2010). https://doi.org/10.1109/TMI.2010.2050897. https://pubmed.ncbi.nlm.nih.gov/20562040/
18. Srivastava, N., Hinton, G., Krizhevsky, A., Salakhutdinov, R.: Dropout: a simple way to prevent neural networks from overfitting. Technical report 56 (2014)
19. Zhao, F., et al.: Harmonization of infant cortical thickness using surface-to-surface cycle-consistent adversarial networks. In: Shen, D., et al. (eds.) MICCAI 2019. LNCS, vol. 11767, pp. 475–483. Springer, Cham (2019). https://doi.org/10.1007/978-3-030-32251-9_52
20. Zhu, J.Y., Park, T., Isola, P., Efros, A.A.: Unpaired image-to-image translation using cycle-consistent adversarial networks. In: Proceedings of the IEEE International Conference on Computer Vision 2017, October, pp. 2242–2251 (2017)

Transfer Learning in Optical Microscopy

Martin Kozlovský(✉), David Wiesner, and David Svoboda

Faculty of Informatics, Masaryk University, Brno, Czech Republic
kozlovsky@mail.muni.cz
https://cbia.fi.muni.cz/

Abstract. Image synthesis is nowadays a very rapidly evolving branch of deep learning. One of possible applications of image synthesis is an image-to-image translation. There is currently a lot of focus orientated to applications of image translation in medicine, mainly involving translation between different screening techniques. One of other possible use of image translation in medicine and biology is in the task of translation between various imaging techniques in modern microscopy. In this paper, we propose a novel method based on DenseNet architecture and we compare it with Pix2Pix model in the task of translation from images imaged using phase-contrast technique to fluorescence images with focus on usability for cell segmentation.

Keywords: Machine learning · GAN · Fluorescence microscopy · Phase-contrast microscopy · Image synthesis

1 Introduction

Living cells are often transparent and therefore not easily visible to human eyes using standard bright-field microscopy. Nonetheless, there exist numerous more advanced microscopy techniques to visualize cells. Two popular and mainly used ones are *phase-contrast microscopy* and *fluorescence microscopy.* These two techniques are very different in terms of price, simplicity of usage, invasivity *etc.* and both have their advantages and disadvantages.

Phase-contrast microscopy uses physical properties of light such as its phase and refractive index to visualize *thickness* of samples. Unlike with some other methods, the cells do not have to be killed, fixed, or stained when using phase-contrast imaging. Due to this noninvasive approach, it can be used for long time cell monitoring without doing any damage to the cells. Unfortunately, phase-contrast does not work well with too thick samples, leading to blurry results. Various other visual artifacts, such as halo effects, are also common [14].

Fluorescence microscopy uses physical principles of fluorescence to image substances. The fluorescent specimen is excited with light which is absorbed and its energy is emitted back a few nanoseconds later, again in a form of light. The excited parts of the specimen are therefore clearly visible against a dark background. This makes fluorescence microscopy very useful to locate and observe even very small substances in a specimen [17].

D. Svoboda et al. (Eds.): SASHIMI 2021, LNCS 12965, pp. 77–86, 2021.
https://doi.org/10.1007/978-3-030-87592-3_8

Nevertheless, fluorescence microscopy has also several disadvantages. One of them is the necessity to stain, dye, or genetically modify the imaged substances. This can interfere with the properties of such substances or introduce visual artifacts and it is also expensive and time consuming. Furthermore, long time exposure to the fluorescent light can cause loss of the fluorescence intensity [15] or, when monitoring living cells, can cause damage to them, resulting in possible deaths of the observed cells [7].

Therefore the ability to synthesize fluorescence images from phase-contrast would be welcomed as it would make possible to only use cheaper and more simple phase-contrast methods and, in some applications, get rid off the time-consuming and expensive fluorescence imaging.

Some work in this area has already been done, for example [1] addresses a similar problem of translation between phase-contrast and fluorescence microscopy with focus to cancer cells segmentation. Our approach differs in biological samples used for training as well as in the method of translation. The authors used Pix2Pix as the deep-learning part whereas we proposed a novel method which outperforms Pix2Pix in this area.

A lot of research has been done in translation between other modalities as well, *e.g.* translation between bright-field and fluorescence microscopy images [8,9], transmitted light and fluorescence microscopy [2,12], or between phase-contrast and differential interference contrast [3].

In this paper, we propose a new method based on DenseNet to perform the translation between phase-contrast and fluorescence microscopy images. The focus is oriented towards the usage in cell segmentation. We compare our method with Pix2Pix, where we show our method produces better results.

2 Data

The dataset is made of paired samples where each pair contains two microscopy images of cells. One image in the pair is imaged using phase-contrast microscopy while the other using fluorescence microscopy with DAPI staining. All images are in grayscale, with the resolution of 1024×1024 pixels and a bit depth of 16 bits. There are 42 pairs of images in total, examples of which can be seen in Fig. 1.

The imaged cells are all human embryonic stem cells (hESC) of the H1 line. This line was discovered in the year 1998 among the first discovered hESC lines ever [6]. It is also one of the most widely used lines in today's research [10].

For the purpose of training and evaluation, the dataset was split into train, validation, and test sets. These sets contain 32, 5, and 5 pairs of samples, respectively.

2.1 Data Augmentation

To deal with the very small size of the dataset, extensive data augmentation was used. Firstly, each image is randomly cropped into a patch of 256×256 pixels.

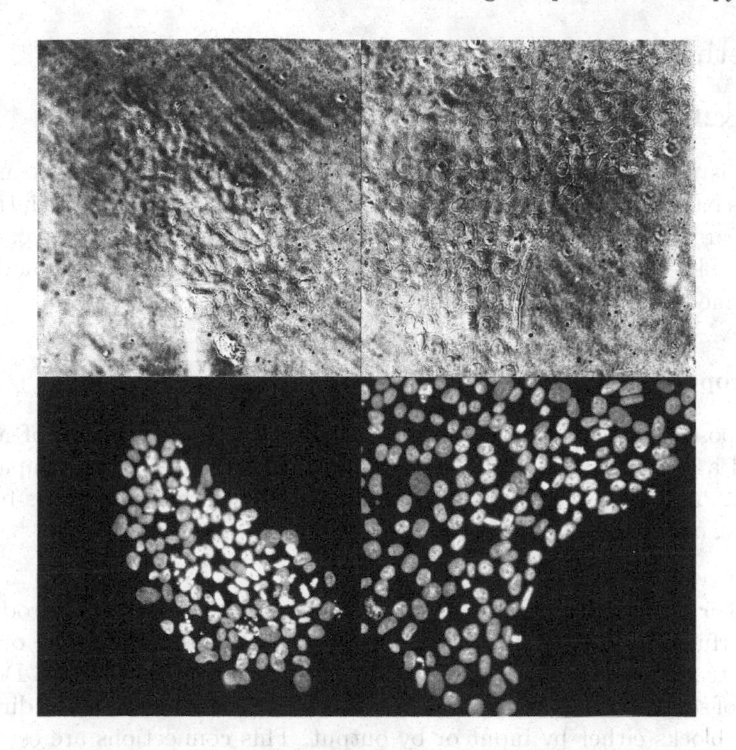

Fig. 1. Example samples drawn from the dataset, with phase-contrast images on top and corresponding fluorescence images below them.

After the cropping, the images are flipped along both the horizontal and the vertical axis with the probability of one half for both directions.

After that, *mixup* [20] augmentation is used. The idea behind this augmentation technique is to produce artificial training samples $\widetilde{x}$ and $\widetilde{y}$ such that

$$\widetilde{x} = \lambda x_i + (1 - \lambda)\, x_j,$$
$$\widetilde{y} = \lambda y_i + (1 - \lambda)\, y_j,$$

where $(x_i,\ y_i;\ x_j,\ y_j)$ are samples independently drawn from the training data and $\lambda \in [0, 1]$ is a randomly chosen value [20].

Furthermore, elastic distortions are applied, as described by Simard *et al.* [19]. The distortions are performed by firstly generating random displacement fields $\Delta x \in [-1, 1]$ and $\Delta y \in [-1, 1]$. These fields are then convolved with a Gaussian filter of a standard deviation σ and multiplied by a scalar factor of α, which controls the intensity of the transformation [19].

The last part of the augmentation consists of slight random adjustments of contrast and brightness. The size of the augmented images is $256 \times 256 \times 1$.

For the test set to be compatible with the trained models, each image in the set is split into 16 non-overlapping patches and those patches which happen to contain no cells are discarded.

3 Methods

3.1 Pix2Pix

Pix2Pix is a deep-learning framework for general image-to-image translation problems based on Conditional GAN (cGAN) [11] architecture. It was introduced in the year 2018 by Isola *et al.* and it was designed to be used universally for various tasks in image translation [5]. In this paper we use this model as a baseline model for its simplicity and good performance.

3.2 Proposed Method

The proposed method is modeled as cGAN, therefore it consists of a generator and a discriminator. For the sake of simplicity and better comparability with Pix2Pix, we used the very same discriminator as well as the loss function. Therefore, only the generator differs.

Generator. The generator is based on DenseNet architecture introduced in 2018 by Huang *et al.* [4]. This architecture was proposed to be one of several methods to solve *vanishing gradient* problem in deep neural networks. DenseNet consists of several convolutional blocks where each block is connected directly to all other blocks either by input or by output. This connections are established as concatenations [4].

The generator contains several sub-networks, which are implemented as variations of U-Net architecture, originally introduced by Ronneberger *et al.* [16]. These specific variations consist of 4 downsampling and upsampling blocks. In each block, there is one convolutional/deconvolutional layer, followed by batch normalization, dropout, and activation function. In the downsampling layers, LeakyReLU activation (with the slope of 0.2) is used; during the upsampling, ReLU activation is used. The filter sizes are 64-128-256-512.

Furthermore, each U-Net contains skip connections in a form of a concatenation between each i and $n - i$ layer, where n is the total number of layers.

The whole generator is then modeled in a similar fashion as DenseNet. The convolutional blocks of DenseNet are only substituted by whole U-Nets. The main idea stays the same, each sub-network is directly connected with all the others. In addition, each sub-network gets to see the original input image again. The schemata of both the architectures are shown in Fig. 2.

By providing the intermediate networks with the original input, the networks have the possibility to enhance some parts of the image without the worry of losing information from the input. The whole chain of the networks should therefore work by gradually extracting information from the input image while having access to the results of all the previous networks. The last sub-network then merges all of the accumulated results throughout the network into one final image.

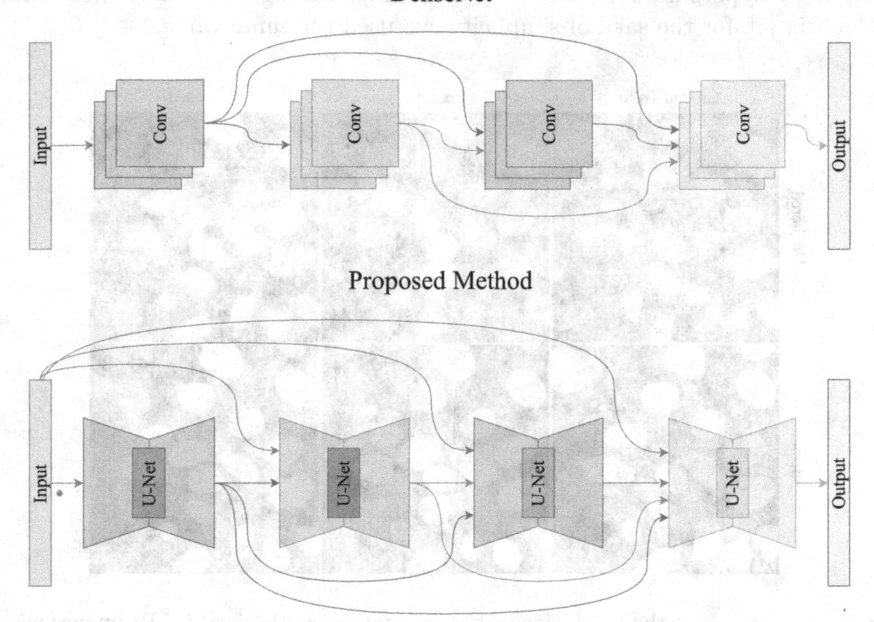

Fig. 2. Comparison of DenseNet and the proposed method. In DenseNet, the input to each block (blocks are denoted using different colors) of convolutions consists of a concatenation of all the outputs from previous blocks. This allows for a more direct flow of gradient from the output layers through the network [4]. The architecture of the proposed model is very similar, the convolutional blocks are replaced by U-Nets. Each U-Net in addition gets to see the input image again. (Color figure online)

Objective. The loss function for cGAN alone can be expressed as [11]:

$$\mathcal{L}_{cGAN}(\mathcal{G}, \mathcal{D}) = \mathbb{E}_{x,y}[\log \mathcal{D}(x, y)] + \mathbb{E}_{x,z}[\log(1 - \mathcal{D}(x, \mathcal{G}(x, z)))], \tag{1}$$

where x is the input, y is the ground truth and z is a vector of random noise. $\mathcal{G}$ and $\mathcal{D}$ denote the generator and the discriminator part, respectively.

Furthermore, it was found to be beneficial to mix the simple cGAN objective with some traditional distance measure loss [13]. Therefore, the loss function for the generator part is extended by L_1 distance between the real and the generated images:

$$\mathcal{L}_{L_1}(\mathcal{G}) = \mathbb{E}_{x,y}[\|y - \mathcal{G}(x)\|_1]. \tag{2}$$

This additional part forces the generator to not only generate pictures which the discriminator can no longer classify as *fake* ones, but also to generate images closer to the ground truth [5].

After incorporating the L_1 loss, the final objective can be written in a form of two player minimax game as [5]:

$$\min_{\mathcal{G}} \max_{\mathcal{D}} V(\mathcal{G}, \mathcal{D}) = \mathcal{L}_{cGAN}(\mathcal{G}, \mathcal{D}) + \lambda \mathcal{L}_{L_1}(\mathcal{G}), \tag{3}$$

where λ is a hyperparameter and its value is empirically set to 100 by the authors of Pix2Pix [5], for the sake of simplicity, we used the same value.

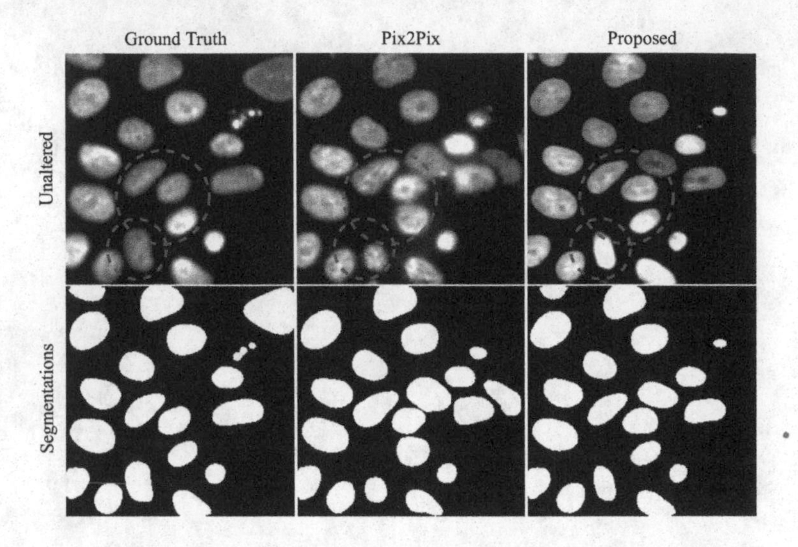

Fig. 3. An example of the limitations when evaluating synthesized DAPI images using only SSIM and RMSE. As to be seen, the ground truth image is slightly blurry. The image produced by Pix2Pix is blurry as well. On the other hand, the image produced by the proposed model is very sharp and the background is much cleaner. The shapes of the cells generated by the proposed model are also more faithful to the ground truth. Nonetheless, the blurriness of the image generated by Pix2Pix causes it to get assigned a better value of both SSIM and RMSE, despite the fact the generated image is clearly worse. For this reason, the evaluation is performed also on the segmentations. As shown in Table 1, when the evaluation is performed this way, our method achieves far better score than Pix2Pix.

Table 1. Metric results for the image used in Fig. 3.

Method	Unaltered		Segmentations	
	SSIM	RMSE	Jaccard	Dice
Pix2Pix	0.7038	0.1832	0.6555	0.7919
Proposed-6	0.2004	0.2731	0.7225	0.8989

4 Experiments

4.1 Parameter Setup

The batch size for training was set to 8. All the models were trained using Adam optimizer with the learning rate 0.0002 and the parameter β_1 set to 0.5. Weights were initialized using normal distribution, with $\mu = 0$, and $\sigma = 0.02$. All the models were trained for 20000 epochs. The proposed architecture was trained in 3 variations, with 4, 6, and 8 intermediate U-Nets. Later in text denoted as Proposed-i, where i is the number of sub-networks.

4.2 Methods of Evaluation

The evaluation was performed not only on the actual ground truth and generated images, but also on their segmentations produced by pretrained StarDist network [18]. The actual generated images are evaluated using SSIM and RMSE. The segmentations are then evaluated using Jaccard Index and Sørensen-Dice Coefficient. The reason for this is to compensate for blurriness of the ground truth images, as demonstrated in Fig. 3.

4.3 Results

Table 2 shows the results of the quantitative measures. In the table, we can see that all variations of the proposed method outperforms Pix2Pix. The overall best one is the Proposed-8 variation which outperforms all the other models by a large margin.

The qualitative results of the translations are shown in Fig. 4. The images generated by Pix2Pix are often not that sharp and the cell surroundings and background tend to be more smudgy. Proposed-4 and Proposed-8 models generate blurry images as well, but their shape is much closer to the ground truth. This is very important for the task of segmentation and shows that the models can have practical use, for example to create a benchmark dataset. Cells in images generated by Proposed-6 model are by far the sharpest, even though the shapes are, based on the metrics, slightly less faithful to the ground truth than the Proposed-8 model.

Table 2. Quantitative results of both the unaltered generated fluorescence images and the segmentations.

Model	Unaltered		Segmentations	
	SSIM	RMSE	Jaccard	Dice
Pix2Pix	0.4461	0.2197	0.6707	0.8029
Proposed-4	0.4581	0.1806	0.7188	0.8364
Proposed-6	0.4424	0.2004	0.7473	0.8554
Proposed-8	0.4968	0.1789	0.7627	0.8654

5 Conclusion

In this paper we proposed a cGAN architecture with DenseNet oriented generator to perform mapping between phase-contrast and fluorescence microscopy images. We compared results of several variations of the proposed method with Pix2Pix, and we showed the images generated by all the proposed methods are more accurate in terms of cell shapes and their general look is also more similar to the ground truth. The Proposed-6 model in addition generates very sharp predictions compared to the other models. All of this makes our method more fitting to be used for generating cell segmentations from phase-contrast images than Pix2Pix.

For future work it can be further experimented with the architecture in terms of the number of sub-networks or also with the depth of the individual U-Nets.

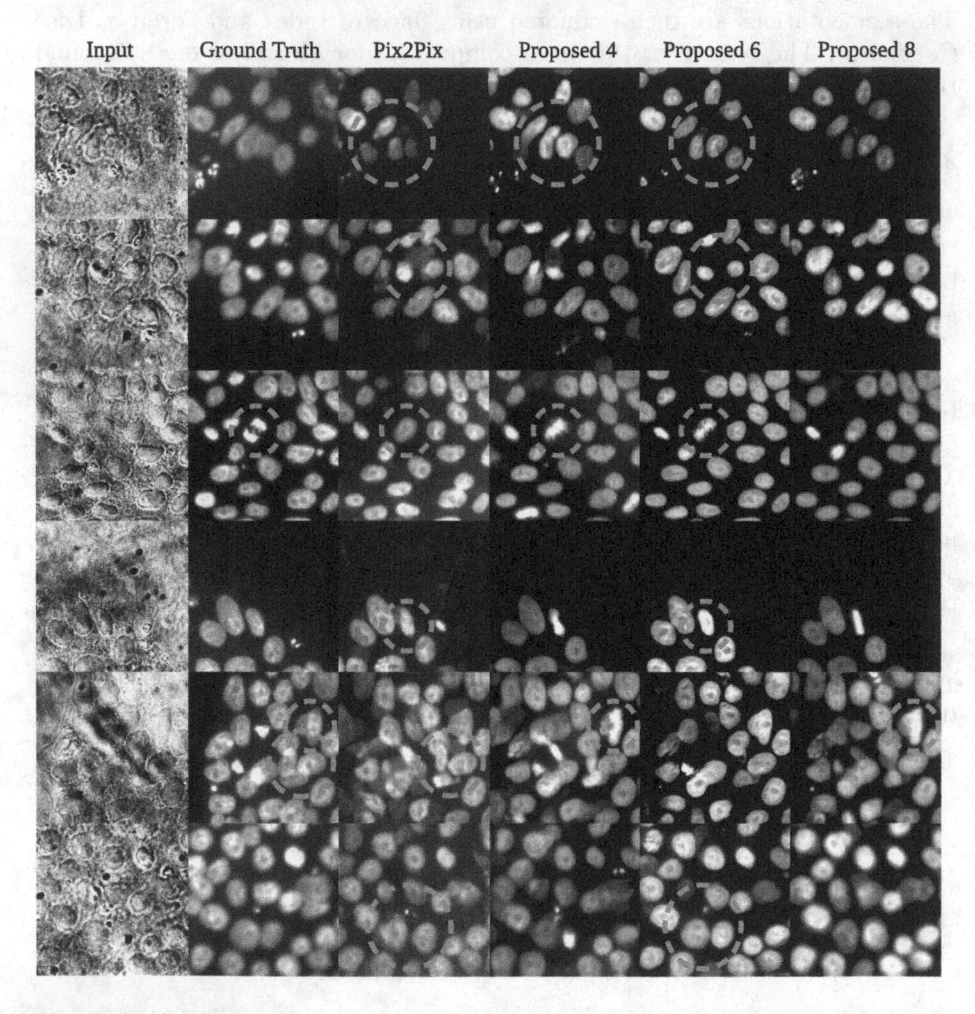

Fig. 4. Qualitative evaluation of the results. All variations of the proposed method generate cells which shapes are more faithful to the ground truth. In the third example, Proposed-4 and 6 variations were able to at least partially spot the presence of mitosis, unlike Pix2Pix, which reconstructed mitosis as a regular cell. In the fifth example, the 8 and 4 variations were able to in some cases correctly generate overlapping cells without merging them together into one homogeneous shape.

References

1. Aida, S., Okugawa, J., Fujisaka, S., Kasai, T., Kameda, H., Sugiyama, T.: Deep learning of cancer stem cell morphology using conditional generative adversarial networks. Biomolecules **10**(6), 931 (2020)
2. Christiansen, E.M., et al.: In silico labeling: predicting fluorescent labels in unlabeled images. Cell **173**(3), 792–803.e19 (2018)
3. Han, L., Yin, Z.: Transferring microscopy image modalities with conditional generative adversarial networks. In: 2017 IEEE Conference on Computer Vision and Pattern Recognition Workshops (CVPRW), pp. 851–859 (2017)
4. Huang, G., Liu, Z., van der Maaten, L., Weinberger, K.Q.: Densely connected convolutional networks (2018)
5. Isola, P., Zhu, J.Y., Zhou, T., Efros, A.A.: Image-to-image translation with conditional adversarial networks (2018)
6. Itskovitz-Eldor, J.: 20th anniversary of isolation of human embryonic stem cells: a personal perspective. Stem Cell Rep. **10**(5), 1439–1441 (2018). May
7. Landry, S., McGhee, P.L., Girardin, R.J., Keeler, W.J.: Monitoring live cell viability: comparative study of fluorescence, oblique incidence reflection and phase contrast microscopy imaging techniques. Opt. Express **12**(23), 5754–5759 (2004)
8. Lee, G., Oh, J.W., Her, N.G., Jeong, W.K.: DeepHCS++: bright-field to fluorescence microscopy image conversion using multi-task learning with adversarial losses for label-free high-content screening. Med. Image Anal. **70**, 101995 (2021)
9. Lee, G., Oh, J.-W., Kang, M.-S., Her, N.-G., Kim, M.-H., Jeong, W.-K.: DeepHCS: bright-field to fluorescence microscopy image conversion using deep learning for label-free high-content screening. In: Frangi, A.F., Schnabel, J.A., Davatzikos, C., Alberola-López, C., Fichtinger, G. (eds.) MICCAI 2018. LNCS, vol. 11071, pp. 335–343. Springer, Cham (2018). https://doi.org/10.1007/978-3-030-00934-2_38
10. Löser, P., Schirm, J., Guhr, A., Wobus, A.M., Kurtz, A.: Human embryonic stem cell lines and their use in international research. Stem Cells **28**, 240–246 (2009)
11. Mirza, M., Osindero, S.: Conditional generative adversarial nets (2014)
12. Ounkomol, C., Seshamani, S., Maleckar, M.M., Collman, F., Johnson, G.R.: Label-free prediction of three-dimensional fluorescence images from transmitted-light microscopy. Nat. Methods **15**(11), 917–920 (2018)
13. Pathak, D., Krahenbuhl, P., Donahue, J., Darrell, T., Efros, A.A.: Context encoders: feature learning by inpainting (2016)
14. Polic, R.: Phase-contrast microscopy principle and applications in materials sciences, May 2020
15. Robson, A.L., et al.: Advantages and limitations of current imaging techniques for characterizing liposome morphology. Front. Pharmacol. **9**, 80 (2018)
16. Ronneberger, O., Fischer, P., Brox, T.: U-Net: convolutional networks for biomedical image segmentation. In: Navab, N., Hornegger, J., Wells, W.M., Frangi, A.F. (eds.) MICCAI 2015. LNCS, vol. 9351, pp. 234–241. Springer, Cham (2015). https://doi.org/10.1007/978-3-319-24574-4_28
17. Rost, F.: Fluorescence microscopy, applications. In: Lindon, J.C., Tranter, G.E., Koppenaal, D.W. (eds.) Encyclopedia of Spectroscopy and Spectrometry, 3rd edn, pp. 627–631. Academic Press, Oxford (2017)
18. Schmidt, U., Weigert, M., Broaddus, C., Myers, G.: Cell detection with star-convex polygons. In: Frangi, A.F., Schnabel, J.A., Davatzikos, C., Alberola-López, C., Fichtinger, G. (eds.) MICCAI 2018. LNCS, vol. 11071, pp. 265–273. Springer, Cham (2018). https://doi.org/10.1007/978-3-030-00934-2_30

19. Simard, P.Y., Steinkraus, D., Platt, J.C.: Best practices for convolutional neural networks applied to visual document analysis. In: Proceedings of the Seventh International Conference on Document Analysis and Recognition, pp. 958–963 (2003)
20. Zhang, H., Cisse, M., Dauphin, Y.N., Lopez-Paz, D.: mixup: beyond empirical risk minimization (2018)

X-ray Synthesis Based on Triangular Mesh Models Using GPU-Accelerated Ray Tracing for Multi-modal Breast Image Registration

J. Maul(✉), S. Said, N. Ruiter, and T. Hopp(✉)

Karlsruhe Institute of Technology, Institute for Data Processing and Electronics, Karlsruhe, Germany
{Johannes.Maul,Torsten.Hopp}@kit.edu

Abstract. For image registration of breast MRI and X-ray mammography we apply detailed biomechanical models. Synthesizing X-ray mammograms from these models is an important processing step for optimizing registration parameters and deriving images for multi-modal diagnosis. A fast computation time for creating synthetic images is essential to enable a clinically relevant application. In this paper we present a method to create synthetic X-ray attenuation images with an hardware-optimized ray tracing algorithm on recent graphics processing units' (GPU) ray tracing (RT) cores. The ray tracing algorithm is able to calculate the attenuation of the X-rays by tracing through a triangular polygon-mesh. We use the Vulkan API, which enables access to RT cores. One frame for a triangle mesh with over 5 million triangles in the mesh and a detector resolution of 1080×1080 can be calculated and transferred to and from the GPU in about 0.76 s on NVidia RTX 2070 Super GPU. Calculation duration of an interactive application without the transfer overhead allows real time application with more than 30 frames per second (fps) even for very large polygon models. The presented method is able to calculate synthetic X-ray images in a short time and has the potential for real-time applications. Also it is the very first implementation using RT cores for this purpose. The toolbox will be available as an open source.

Keywords: X-ray simulation · Ray tracing · GPU · Triangular mesh · Multi-modal image registration · Bio-mechanical Model

1 Background

Image registration is a crucial step for image analysis because valuable information from different images can be combined. The goal of image registration is to find corresponding locations in two or more images and combine the information of these images [9]. Therefore, the accurate integration of the information from two or more images is important for the quality of the result [7].

D. Svoboda et al. (Eds.): SASHIMI 2021, LNCS 12965, pp. 87–96, 2021.
https://doi.org/10.1007/978-3-030-87592-3_9

For early breast cancer diagnosis often multiple modalities are applied. To combine those modalities an image registration is important due to the considerably different patient positioning, dimensionality and deformation state of the breast. To tackle these huge non-linear deformations we develop image registration methods based on biomechanical models of the breast, e.g. for combining the diagnostic values of Magnetic Resonance Images (MRI) and X-ray mammograms. For this purpose a synthetic X-ray image has to be generated from the biomechanical model, which is represented by a polygoal mesh, in order to compare it to the real mammogram for iterative optimization of registration parameters and final combined diagnosis [4,5,10]. For future application in which e.g. biomechanical model generation may be computed on GPUs, the computation time is essential to enable a clinically relevant application.

Beside our own application, the demand for synthetic X-ray images is high because it not just enables image registration but also a wide variety of medical imaging applications such as evaluation of CT reconstruction algorithms, and can furthermore be used e.g. in the field of non-destructive testing [3].

The geometry of (biomechanical) anatomical models are often described with polygon meshes. Typically the polygons are triangles, rectangles or hexagons. In our image registration method, biomechanical models are automatically created from segmented MRI volumes using the iso2mesh [1] toolbox which is based on tetgen [11] and Computational Geometry Algorithms Library (CGAL) [8]. It produces tetrahedral meshes which are subsequently applied for a deformation simulation with FEM. These polygon meshes allow the use of ray tracing algorithms for calculation of the interaction of X-rays with the tissue in order to create synthetic X-ray images.

The available toolboxes for the calculation of synthetic X-ray images are mostly based on volume images using ray casting, thereby requiring an additional processing step to convert the mesh into a voxel volume in case of image registration application [13].

Few implementations exist with respect to X-ray simulation based on polygon meshes. Due to the similarity to ray tracing applications in computer vision, these methods are typically well suited for parallel processing on GPUs. An example for such a ray tracing based algorithm on GPU and with L-Buffers is presented in [12]. The computation time demonstrated is approx. 10 ms for a detector resolution of 1024×768 pixels and a mesh with 11,102 triangles.

An important motivating factor for this work was the development of ray tracing (RT) cores in recent graphics card generations. Because ray tracing is often considered as too computationally intensive, RT cores accelerate the ray tracing principle with hardware-optimized intersection calculation. To our best knowledge, there is no toolbox available using the RT cores of recent GPU generations for X-ray simulations in medical application. This work therefore aims at incorporating the recent advances in GPU computing for fast generation of synthetic X-ray images from potentially very large polygon meshes. In order to combine the advantages of GPU accelerated computing and prototyping in scripting languages, such as MATLAB, we made use of the Vulkan API [14] for implementation of the ray tracing functionality and furthermore developed a MATLAB wrapper to evaluate the results and enable further processing.

2 Methods

2.1 Basic Principle

X-rays traveling through a material are attenuated depending on the material characteristics. The received X-ray intensity at a detector depends on the initial intensity of the X-ray source, the absorption property of the material and the distance it travels trough the material. This relationship can be expressed by the Beer-Lambert-Law, which is the basic physical principle we use for our application. The Beer-Lambert-Law is defined as

$$I = I_0 \exp\left(-\int \mu(x_n)dx\right) \tag{1}$$

where I_0 represents the initial intensity of the X-ray emitted at the X-ray source, I_n represents the intensity received at a detector position n. Nowadays digital detectors are mainly used for data acquisition, which discretize the field of view into pixels. x_n expresses a position on the ray connecting source and detector. The function $\mu(x_n)$ describes the corresponding attenuation coefficient at x_1. The integral integrates over the traversed part in the material.

Assuming locally isotropic materials, the integral can be discretized as the sum of the product of the partial distances a ray travels through one material and the material's X-ray attenuation coefficient.

To determine the partial distances, which a ray travels through the respective materials in a polygon mesh, intersections of the ray and the polygons need to be calculated, which is a classical ray tracing problem known from computer vision. Once intersection points are known, the partial distance can be calculated by the Euclidean distance between the intersection points. Expressing one ray by a directional vector $\overrightarrow{d}$ from source to detector, all intersection points can be described by $t \cdot \overrightarrow{d}$ where t is a scalar describing the proportional distance from the source. Following this approach the Beer-Lambert law can be re-written as

$$I_n = I_0 \exp\left(-\sum_{k=1}^{(m-1)} (\|(t_{k+1} - t_k)\overrightarrow{d}\| \, \mu(x_{m,n}))\right) \tag{2}$$

where m is the number of intersections between ray and polygon mesh and t_k is the relative distance from the source to a specific intersection. An visual example is given in Fig. 2.

In our implementation we define a point source for the X-rays. Rays originating from the source are tracked through the polygon mesh using ray tracing until they hit a detector pixel. For this purpose a directional vector from the source to a detector is calculated. Refraction is currently neglected and hence the tracing is performed along straight lines. The expansion of an X-ray is not considered. Detector pixels are modeled as point detectors with an ideal transfer function.

The polygon mesh represents the surfaces of different tissues. Multiple surfaces can be interlaced. In our data structure we encode the material type by assigning region IDs to every polygon, which together with a lookup table are able to identify the attenuation coefficient for the respective material.

2.2 Implementation on Special Purpose Ray Tracing Units of GPUs

For implementing the algorithms on the special purpose RT cores of GPUs we use the Vulkan API, which is a standardized API for GPU applications provided by the Khronos group [14]. Vulkan executes commands on driver level and allows cross-platform implementation. To achieve compatibility with different GPU brands, it is possible to let Vulkan take over the execution of functions called by the user.

Vulkan provides special data structures for ray tracing problems. The polygon mesh is represented by two entities (nodes and faces). This data structure is also known as face-vertex mesh. These two entities can be forwarded to Vulkan, which creates an acceleration structure (AS). The node entity contains the coordinates of the vertices of a polygon. The faces entity holds the indices of vertices, which are connected to polygons. To use hardware accelerated intersection calculation provided by RT cores, the polygon shapes are restricted to triangles.

The implementation is based on the well known examples provided by Willems [15]. For generating and handling rays, three shaders are used. The "raygen shader" calculates the ray direction $\overrightarrow{d}$ and defines the parameter of the ray. After that the raygen shader starts the ray tracing with an Vulkan API call *traceNV*. The Vulkan library then executes a selected shader whenever an intersection between the ray and the AS is detected. We use the "anyhit shader" to acquire all information about the intersection such as the material type $\mu(x)$ and the distance t, calculated with full precision (32 bit). After all intersections are detected or after a ray does not further intersect with another triangle, the "miss shader" will be called by Vulkan. In this shader we implement the attenuation calculation by sorting the intersections and summing up the product of distances and attenuation coefficients according to the Beer-Lambert law given in equation Eq. 2.

Figure 1 illustrates the implementation of shaders for the described functionality. The construction of the AS and the execution of the shaders are performed by the Vulkan library. In order to start the ray tracing shaders, the host calls the *QueueSubmit* function.

We implemented the ray tracing principle with Vulkan in a headless-mode, i.e. the application does not display the resulting images. Instead we integrated the method into the MATLAB scripting language via a MEX interface.

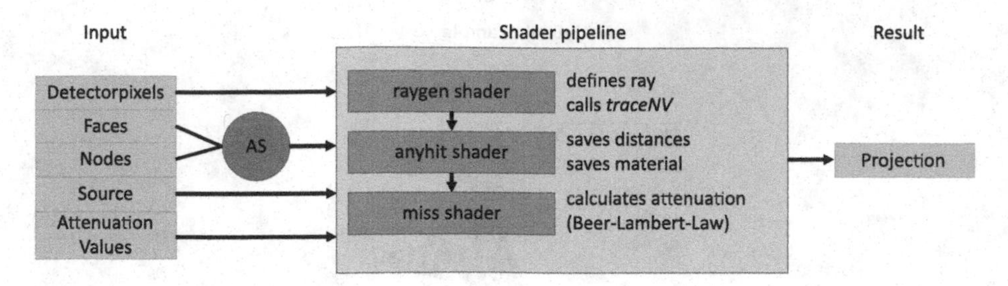

Fig. 1. Simplified X-ray calculation. The parameters are passed via a MEX interface, the AS is built by Vulkan. Then the shaders get executed, which calculates the attenuation with the Beer-Lamber-Law. The result is an synthetic X-ray projection.

3 Results

3.1 Validation of Attenuation Calculation

In order to validate the implementation we calculated the attenuation of an analytical example. For this purpose, triangles belonging to different tissue types are placed perpendicular to a ray. We define one detector pixel and source position. Using the analytic positions of source, detector and triangles makes it possible to calculate the analytic attenuation and test the result of the implementation against it (Fig. 2).

The deviation is in the range of the data type precision for every tested analytical example. For the example given in Fig. 2 the relative error of the computed attenuation was $5.7 \times 10^{-8}\%$. The analytic example can be recreated and further tested as a part of the open source toolbox.

In Fig. 3 the synthetic X-ray images generated by the presented method for four examples are shown for visual assessment. The breast example shows small artifacts at the bottom edge caused by holes in the underlying mesh and not by the toolbox.

To demonstrate the application for image registration based on biomechanical models, Fig. 4 shows a synthetic X-ray image based on the deformed biomechanical model compared to an real X-ray mammogram. We can see, that the shape of the breast and the inner structures look similar, even though the fine structures are not visible in the synthetic X-ray image due to limited resolution of the underlying MRI data.

3.2 Performance Evaluation

To evaluate the computational performance we created several examples with a varying number of triangles and a varying resolution of the detector. Both parameters have a direct effect on the calculation time.

Table 1 presents the different properties of the examples and the execution times for one image. Both the computation time with and without data transfer to and from the GPU are illustrated.

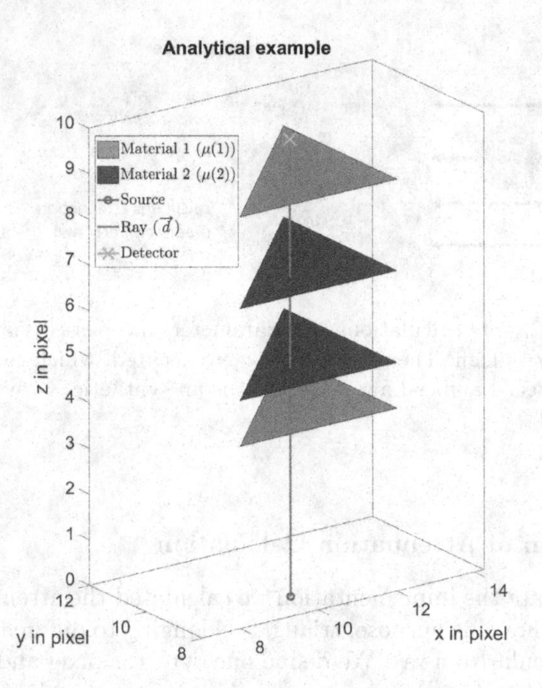

Fig. 2. Structure of the analytical example to determine the accuracy with two layers. The blue circle represents the source. The triangles are at the material surface in a mesh and the different colours represent different material types. The orange line represents the ray and the end is e.g. at the detector plate or at an defined maximum length. (Color figure online)

Thereby we created two scenarios: (1) Direct application with MATLAB MEX interface for calculating one synthetic X-ray image, (2) interactive application in which only the rendering for the X-ray synthesis on the GPU is considered and data transfer to/from the GPU are neglected. This second scenario may be thought of e.g. a real time renderer with varying source position while the polygon mesh data is not changed.

All calculations are performed on a NVidia RTX 2070 Super GPU on a Windows workstation PC.

A basic observation from Table 1 is the considerable overhead of the data transfer to and from the GPU by comparing the calculation times of scenario 1 and scenario 2. With increasing number of triangles the calculation time in scenario 2 increases. Furthermore it can be noticed that e.g. the "veins" example with considerably less triangles than the "breast" example, but more pixels on the detector plane takes longer to compute for scenario 1.

To further investigate the dependencies of the computation time, the geometry used in the "breast" example was taken as a basis to create polygon meshes with different number of triangles. Additionally the detector resolution was varied. Figure 5 shows the results of this analysis in frames per second (fps).

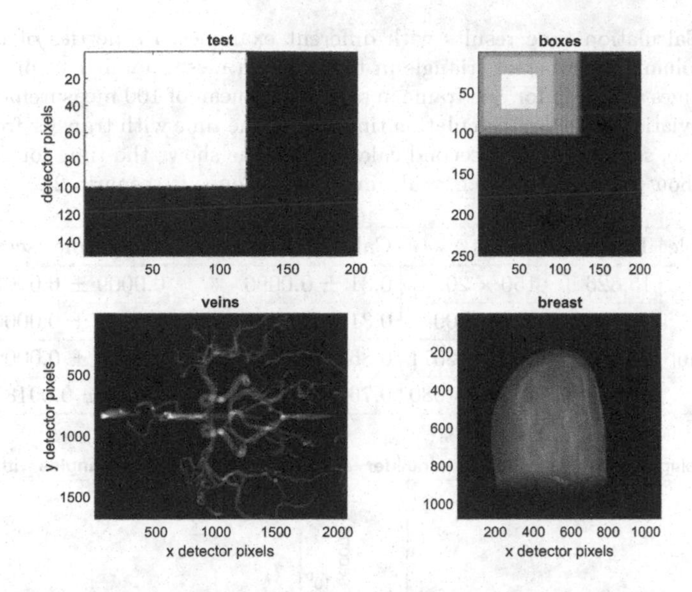

Fig. 3. Exemplary results of the X-ray simulation for the examples from Table 1. The "test" example models a simple plate made from one material. The "boxes" example models three interleaved boxes of different material. The "veins" example was provided by [6] and displays a part of the veins in the human body. The "breast" example models the geometry of a breast based on a segmented MRI scan as typically used in our biomechanical model based image registration.

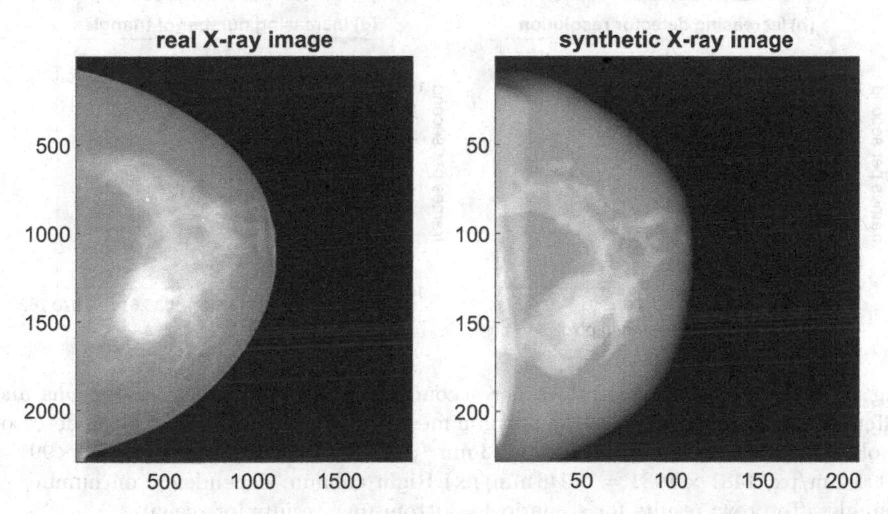

Fig. 4. Comparison between real and synthetic X-ray image.

As we can see, the detector resolution has a stronger influence on the computational performance than the number of triangles, because the higher resolution leads to more possible intersections as more rays are generated and more hits are detected. The difference in fps is small up to the 113,000 triangle example

Table 1. Calculation time results with different examples. Properties of the examples (first column), number of triangle in the polygon mesh, number of pixels on the detector. Times are given for one frame in seconds as mean of 100 measurements ± the standard deviation. The first calculation time shows the time with transfer from and to MATLAB (i.e. scenario 1), the second calculation time shows the time for computing 1 frame without data transfer, e.g. real time application (i.e. scenario 2).

Example	Triangles	Pixels	Calc. time (scen. 1)	Calc. time (scen. 2)
Test	15,626	150 × 200	0.31 ± 0.0096	0.0009 ± 0.000087
Boxes	47,580	250 × 200	0.31 ± 0.0089	0.0014 ± 0.000088
Veins [6]	215,728	1681 × 2071	0.86 ± 0.0117	0.0041 ± 0.000031
Breast	5,185,459	1080 × 1080	0.76 ± 0.0117	0.025 ± 0.0018

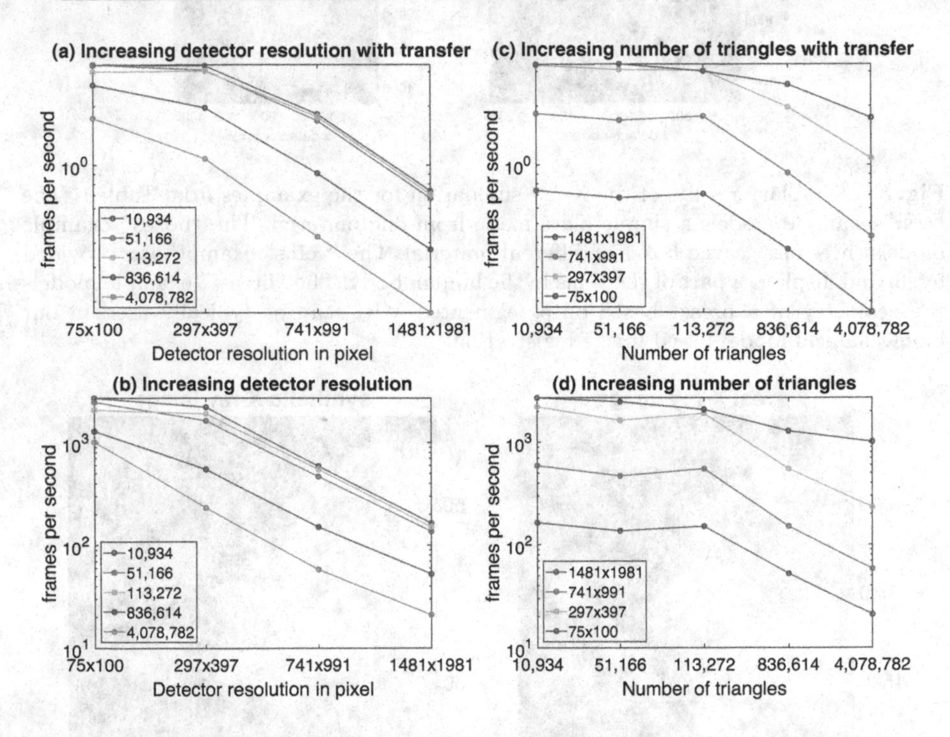

Fig. 5. Rendering times in frames per second for different detector resolutions and different amount of triangles in the polygon mesh. Left column: dependency on detector resolution. (Resolutions: $75 \times 100 = 0.911$ mm/px, $297 \times 397 = 0.23$ mm/px, $741 \times 991 = 0.091$ mm/px, $1481 \times 1981 = 0.046$ mm/px) Right column: dependency on number of triangles. Top row: results for scenario 1. Bottom row: results for scenario 2.

for both scenario 1 and 2, while a constant decrease in fps can be observed for increasing detector resolution.

For all examples, the overall fps is considerably higher without data transfer. In scenario 1, for large problems (around 4 million triangles, detector resolution

1481 × 1981 pixels) the fps is nearly 1, while small problems (around 11,000 triangles, detector resolution 75 × 100 pixels) reach about 30 fps. Scenario 2 reaches up to 21 fps for the large problem and up to 2700 fps for the small one.

Deeper analysis of the differences between scenario 1 and scenario 2 showed that the data transfer and the initialization of the Vulkan library need more than 95% of the computation time of one frame in scenario 1. As these steps have to be done only once for a real time application, this explains the huge time differences between scenario 1 and scenario 2.

4 Discussion and Conclusion

In this paper we presented a method for synthesizing X-ray attenuation images based on polygon meshes using ray tracing. To our best knowledge, the provided toolbox is the first example using RT cores for such purposes as other toolboxes concentrate on memory-management [12,13] or volume based methods [2]. Because this is the first approach to use RT cores to generate synthesizing X-ray attenuation images, this paper focuses on the implementation. In future, comparison with other proposals will be made in more detail. With the MEX interface an easy access from a high level scripting language, which is often used in scientific applications, is possible.

The main purpose of this method is to accelerate the biomechanical model based image registration of MR images and X-ray mammograms in future. While this has been easily achieved compared to the former applied method using ray casting through a voxel grid [4], the real time potential of the method is even higher and may open up new fields of application. The method has limitations, like a straight ray approximation, negligence of refraction and considering a monochromatic case only. However these limitations may be overcome in future, because e.g. refraction is common in ray tracing problems in computer vision and may as such be easily introduced in the algorithm.

The computational performance reached by the current implementation is very promising. Further optimization may concern the data transfer to and from the GPU as well as potentially reducing the overhead by the Vulkan library. Furthermore in the current implementation the execution time of the miss shader needs between 30% and 50% of the total rendering time for one frame. This may be optimized by replacing the current sorting method of intersections with e.g. the approach presented in [12].

Even though the resulting toolbox was originally developed for synthesizing X-ray images for our model-based image registration, the performance of the GPUs has the potential, that the simulator can be used as a real time application for other purposes than registration, as we demonstrate in the result section.

The presented methods have been assembled in a toolbox such that it can be used in a wide field of biomedical applications. To achieve this goal, we share this toolbox as open source project including all examples presented in this paper, which can be accessed at the following web page:

https://git.scc.kit.edu/dach/raytracingx-rayprojectionsimulator

References

1. Fang, Q., Boas, D.A.: Tetrahedral mesh generation from volumetric binary and grayscale images. In: 2009 IEEE ISBI: From Nano to Macro, pp. 1142–1145 (2009). https://doi.org/10.1109/ISBI.2009.5193259
2. Folkerts, M., Jia, X., Gu, X., Choi, D., Majumdar, A., Jiang, S.: MO-FF-A4-05: implementation and evaluation of various DRR algorithms on GPU. Med. Phys. **37**(6Part6), 3367 (2010). https://doi.org/10.1118/1.3469159
3. Hanke, R., Fuchs, T., Uhlmann, N.: X-ray based methods for non-destructive testing and material characterization. Nucl. Instrum. Methods Phys. Res. **591**(1), 14–18 (2008). https://doi.org/10.1016/j.nima.2008.03.016
4. Hopp, T., et al.: Automatic multimodal 2D/3D breast image registration using biomechanical FEM models and intensity-based optimization. Med. Image Anal. **17**(2), 209–218 (2013). https://doi.org/10.1016/j.media.2012.10.003
5. Hopp, T., Baltzer, P., Dietzel, M., Kaiser, W.A., Ruiter, N.V.: 2D/3D image fusion of X-ray mammograms with breast MRI: visualizing dynamic contrast enhancement in mammograms. IJCARS **7**(3), 339–348 (2012). https://doi.org/10.1007/s11548-011-0623-z
6. Lickteig, S.: Segmentierung & Visualisierung von Blutgefößen, Seminar paper, University of Cooperative Education, Karlsruhe (2021)
7. Oliveira, F.P.M., Tavares, J.M.R.S.: Medical image registration: a review. CMBBE **17**(2), 73–93 (2014). https://doi.org/10.1080/10255842.2012.670855
8. Rineau, L., Yvinec, M.: 3D Surface Mesh Generation (2008). https://doc.cgal.org/latest/Surface_mesher/index.html. Accessed 26 July 2021
9. Rueckert, D., Schnabel, J.A.: Medical image registration. In: Deserno, T.M. (ed.) Biomedical Image Processing. BIOMEDICAL, pp. 131–154. Springer, Heidelberg (2011). https://doi.org/10.1007/978-3-642-15816-2_5
10. Said, S., Clauser, P., Ruiter, N.V., Baltzer, P.A.T., Hopp, T.: Image registration between MRI and spot mammograms for X-ray guided stereotactic breast biopsy: preliminary results. In: Linte, C.A., Siewerdsen, J.H. (eds.) Medical Imaging 2021: Image-Guided Procedures, Robotic Interventions, and Modeling, p. 45. SPIE (15022021–20022021). https://doi.org/10.1117/12.2581820
11. Si, H., Gärtner, K.: Meshing piecewise linear complexes by constrained delaunay tetrahedralizations. In: Hanks, B.W. (ed.) Proceedings of the 14th International Meshing Roundtable, pp. 147–163. Springer, Heidelberg (2005). https://doi.org/10.1007/3-540-29090-7_9
12. Vidal, F.P., Garnier, M., Freud, N., Létang, J.M., John, N.W.: Simulation of X-ray attenuation on the GPU. In: Tang, W., Collomosse, J. (eds.) Theory and Practice of Computer Graphics. The Eurographics Association (2009). https://doi.org/10.2312/LocalChapterEvents/TPCG/TPCG09/025-032
13. Vidal, F.P., Villard, P.F.: Development and validation of real-time simulation of X-ray imaging with respiratory motion. CMIG **49**, 1–15 (2016). https://doi.org/10.1016/j.compmedimag.2015.12.002
14. Vulkan: Home—Vulkan—Cross platform 3D Graphics (2021). https://www.vulkan.org/. Accessed 20 May 2021
15. Willems, Sascha: SaschaWillems/Vulkan Examples (28052021). https://github.com/SaschaWillems/Vulkan. Accessed 28 May 2021

Application-Oriented Papers

Frozen-to-Paraffin: Categorization of Histological Frozen Sections by the Aid of Paraffin Sections and Generative Adversarial Networks

Michael Gadermayr[1(✉)], Maximilian Tschuchnig[1], Lea Maria Stangassinger[2], Christina Kreutzer[3], Sebastien Couillard-Despres[3], Gertie Janneke Oostingh[2], and Anton Hittmair[4]

[1] Department of Information Technology and Systems Management, Salzburg University of Applied Sciences, Salzburg, Austria
Michael.Gadermayr@fh-salzburg.ac.at

[2] Department of Biomedical Sciences, Salzburg University of Applied Sciences, Salzburg, Austria

[3] Institute of Experimental Neuroregeneration, Spinal Cord Injury and Tissue Regeneration Center Salzburg (SCI-TReCS), Paracelsus Medical University, Salzburg, Austria

[4] Department of Pathology and Microbiology, Kardinal Schwarzenberg Klinikum, Schwarzach im Pongau, Austria

Abstract. In contrast to paraffin sections, frozen sections can be quickly generated during surgical interventions. This procedure allows surgeons to wait for histological findings during the intervention to base intra-operative decisions on the outcome of the histology. However, compared to paraffin sections, the quality of frozen sections is typically lower, leading to a higher ratio of miss-classification. In this work, we investigated the effect of the section type on automated decision support approaches for classification of thyroid cancer. This was enabled by a data set consisting of pairs of sections for individual patients. Moreover, we investigated, whether a frozen-to-paraffin translation could help to optimize classification scores. Finally, we propose a specific data augmentation strategy to deal with a small amount of training data and to increase classification accuracy even further.

Keywords: Histology · Frozen sections · Generative adversarial networks · Thyroid cancer · Data augmentation · Whole slide image classification

1 Motivation

Whole slide imaging is capable of effectively digitizing specimen slides, showing both the microscopic detail and the larger context, without any significant manual effort. Visual pathological examination of slides, independent whether they

The original version of this chapter was revised: an error in the affiliation of one of the authors was corrected. The correction to this chapter is available at https://doi.org/10.1007/978-3-030-87592-3_15

D. Svoboda et al. (Eds.): SASHIMI 2021, LNCS 12965, pp. 99–109, 2021.
https://doi.org/10.1007/978-3-030-87592-3_10

are digitized or not, is time-consuming and error prone, due to the large amount of information available in the highly-resolved image data. These facts provide the incentive for the development of automated tools to support pathologists during clinical routine. Recently, automated segmentation [4,6,17], normalization [2] and classification approaches [7] have been proposed, mainly based on state-of-the-art deep learning approaches.

Tissue fixation is the first step in the histological process of tissue preparation. In the case of paraffin sections, the tissues are formalin fixed and subsequently embedded in paraffin as a solid medium both to preserve the tissue and to enable the cutting of thin tissue slices. Paraffin embedding is probably the most commonly used technique (Fig. 1, bottom row) that is compatible with a large variety of staining methods and allows thin sectioning (down to few micrometers) with a high visual quality. So-called frozen sections (Fig. 1, top row), are typically generated during interventions (e.g. cancer resections) in order to achieve information on malignancy as quick as possible, because the preparation time for paraffin sections is typically too long to be used for that purpose. Frozen sections therefore allow surgeons to wait during interventions for the histological examination in order to base further procedures on the outcome. However, compared to paraffin sections, the quality of frozen sections is typically lower, leading to a higher ratio of miss-classification when performed by clinical experts [9,13,14]. The cellular structure can be seen more clearly in paraffin sections, as these are fixed before embedding in paraffin. This is not the case in frozen sections with the result of partly indiscernible or damaged tissue features [11,16].

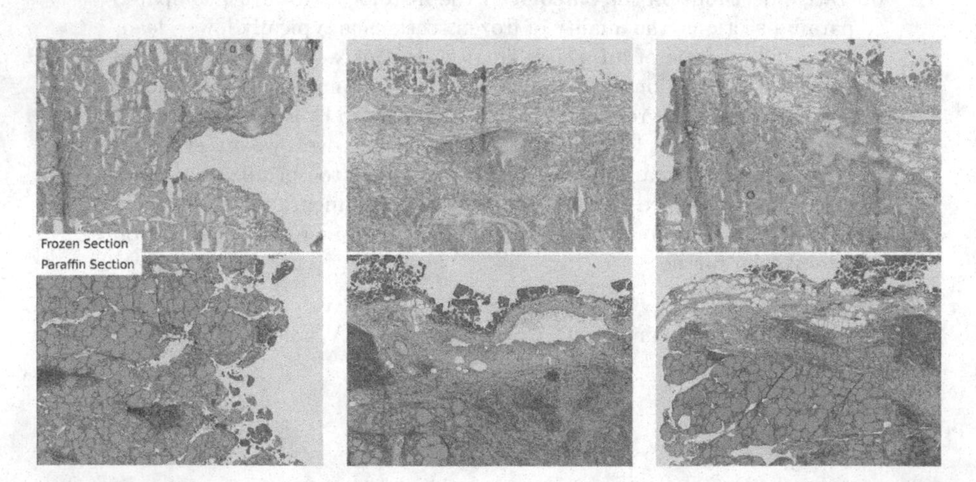

Fig. 1. Semi-corresponding tissue showing frozen sections (top row) and paraffin embedded sections (bottom row). The two preparations were obtained from neighboring tissue (and thereby show similarities).

Generative adversarial networks, recently proved to facilitate a translation between medical imaging domains, such as CT to MRI [18] or translations

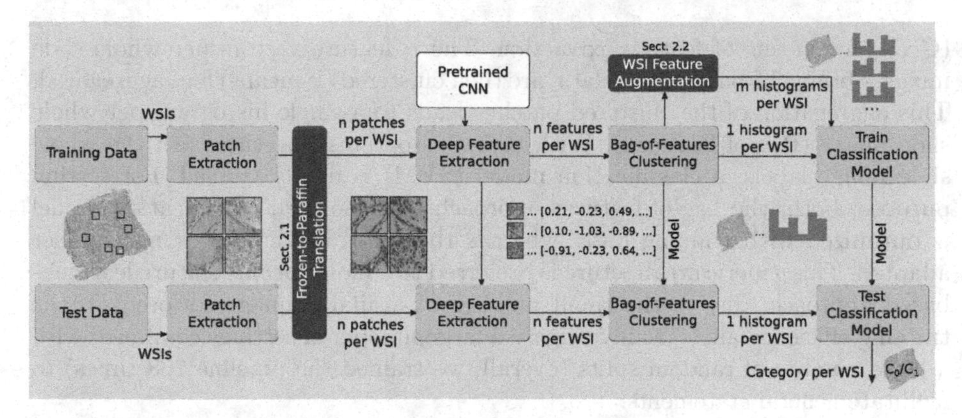

Fig. 2. Outline of the proposed basic pipeline with the proposed additional stages (Frozen-to-Paraffin Translation & WSI Feature Augmentation) for improving accuracy of frozen sections. Input of the pipeline is a set of whole slide images for training and one for testing. Output is a classification label for each test image.

between different stainings [4], based on unpaired image data only. A limitation in this application scenario is often the fact that so-called one-to-many mappings [1] complicate training of cyclic architectures [19,20]. The challenge in many application scenarios is, that a characteristic in one domain often can be mapped to more than one pendent in the other modality. This can lead to conflicts when using pixel-based losses [19,20]. Even more recently, methods-of-resolution were proposed, tackling also highly challenging translation settings by adding an additional latent variable [1,8,12] or by replacing the pixel-based cycle-consistency loss with a feature-based loss [15].

Contribution. The contribution of this work is manifold: Based on a generic fully automated pipeline, we first investigated the effect of the section type on automated decision support approaches for the classification of thyroid cancer. A unique data set showing similar tissue from both modalities, allowed for an unbiased evaluation. Secondly, we investigated, whether a frozen-to-paraffin conversion can help to increase classification scores of frozen sections. To improve the performance in the case of small data even further, a specific data augmentation strategy is proposed.

2 Methods

To investigate the impact of the section type and section-translations on the classification accuracy of automated approaches, we rely on a generic straightforward patch-based whole slide image classification architecture (Fig. 2), similar to [3,7]. As a holistic classification of whole slide images is not effective [7], in a first step patches are randomly extracted in image areas showing mainly tissue. These patches are fed into a pre-trained convolutional neural network

(CNN) for means of feature extraction. The n feature vectors per whole slide image, obtained from training data, are then clustered (k-means) and aggregated. This aggregation of the clustered patches leads to a single histogram per whole slide image (Bag-of-Features). With these histograms, together with the whole slide image labels, a classification model (SVM) is finally trained. For testing purpose, both, the bag-of-features approach and the final classification model is optimized with training data, whereas the pre-trained CNN is not further adapted. This generic architecture is preferred to a more specific feature learning-based approach, due to the small number of available images. Moreover, the training efficiency allows to investigate a large number of settings combined with a large number of random splits (overall, we trained the pipeline 768 times) to facilitate general statements.

We propose two adjustments: firstly, we propose to apply frozen-to-paraffin translation to optimize the quality of frozen sections, by means of a state-of-the-art image-translation approach (Sect. 2.1). Secondly, we propose to augment the available training samples (per whole slide image), by a very efficient sampling strategy (Sect. 2.2). This adjustment allows the simulation of a larger amount of available training data (even though a larger data set would be advantageous) to make more general statements on the effect of image-translation.

2.1 Frozen-to-Paraffin Translation

For frozen-to-paraffin translation, a generative adversarial network based on contrastive learning [15] is employed, which solves the problem of one-to-many mappings of cyclic-architectures [20] by means of a feature based loss.

We denote the domain of frozen sections as F ($f \in F$) and the domain of paraffin sections as P ($p \in P$). We trained the generator G consisting of an encoder G_{enc} and a decoder G_{dec} to perform the mapping $\hat{p} = G_{dec}(G_{enc}(f))$ were $\hat{p}$ represents the virtual paraffin pendant of the original frozen section f. We optimized a loss criterion consisting of a weighted sum of a GAN loss $\mathcal{L}_{GAN}$, a patch similarity loss $\mathcal{L}_{PatchNCE}(G, H, F)$ (NCE stands for noise-contrastive estimation) forcing corresponding patches to share content and an additional regularization term $\mathcal{L}_{PatchNCE}(G, H, P)$. In summary, the loss can be formulated as follows

$$L = \mathcal{L}_{GAN}(G, D, F, P) + \lambda_F \mathcal{L}_{PatchNCE}(G, H, F) + \lambda_G \mathcal{L}_{PatchNCE}(G, H, P), \quad (1)$$

with D being the discriminator, H being a two-layer perceptron and λ_F and λ_G being scalar weights. The patch similarity loss was computed on features level, i.e. it was computed after the encoder part of the generator (G_{enc}) was applied to the image. We utilized a ResNet with nine blocks as the generator, patchGAN [10] as the discriminator, a least squares GAN loss, a batch size of one and Adam as an optimizer with an initial learning rate of 0.002. The model was trained for 40 epochs. λ_F and λ_G were both set to 1.0. For details, we refer to [15]. Finally, we used the Pytorch reference implementation with the default 'cut' setting.

2.2 WSI Feature Augmentation

Due to the rather small amount of available whole slide images, we incorporated feature-augmentation to obtain several features for one whole slide image. In the patch extraction stage, for each original image, 512 patches with a size of 256×256 pixels were randomly (uniformly) selected from regions showing (at least 75%) tissue. In the bag-of-features stage, each patch was assigned to a cluster center. The final feature vector was represented by the distribution of patches with respect to these cluster centers. In the conventional setting, this generates exactly one feature vector for each whole slide image.

To increase the number of feature vectors for training, we reduced the number of sampled patches to a certain ratio (r_{patch}) of the 512 originally extracted patches and repeated this process eight times. Thereby, we achieved a clearly increased number of features (8×) without high computational or manual effort. Obviously, the features created within one whole slide images showed a higher correlation than features between images. However, the motivation for this procedure was provided by the patchy distribution of relevant disease markers. The generation of several random samples can help to properly fill the feature space before training a classification model. In the experimental evaluation, two different sampling strategies (Aug1: $r_{patch} = 75\%$, Aug2: $r_{patch} = 50\%$) were evaluated.

2.3 Data Set

In this work, we aimed at distinguishing different nodular lesions of the thyroid, focusing especially on benign follicular lesions and papillary carcinoma. This differentiation is crucial, due to different treatment options, in particular with respect to the extent of surgical resection of the thyroid gland. The data set consisted of 80 whole slide images overall, i.e. 40 slides were available for each section type. All images were acquired during clinical routine at the Kardinal Schwarzenberg Hospital. They were labeled by an expert pathologist with over 20 years experience. A total of 42 (21 per modality) slides were labeled as papillary carcinoma while 38 (19 per modality) were labeled as benign follicular lesion. During the generation of data, the focus was placed on extracting samples showing similar tissue (for examples, see Fig. 1). Perfect matching was not possible, since the tissue blocks needed to be separated and individually processed for each modality. As a result, two consecutive slides showing the perfect immediately neighboring tissue areas [5] cannot be obtained for this procedure. Nevertheless, the slides showed similar underlying content, which is an important criterion when comparing classification accuracy for both modalities. However, due to the imperfect alignment of the data, paired image translation approaches were not applicable. For frozen sections, fresh tissue was frozen at −15 °C, slides were cut (thickness 5 µm) and stained immediately with hematoxylin and eosin. For paraffin sections, tissue was fixed in 4% phosphate-buffered formalin for 24 h. Subsequently formalin fixed paraffin embedded tissue was cut (thickness 2 µm) and stained with hematoxylin and eosin. Images were digitized with an Olympus

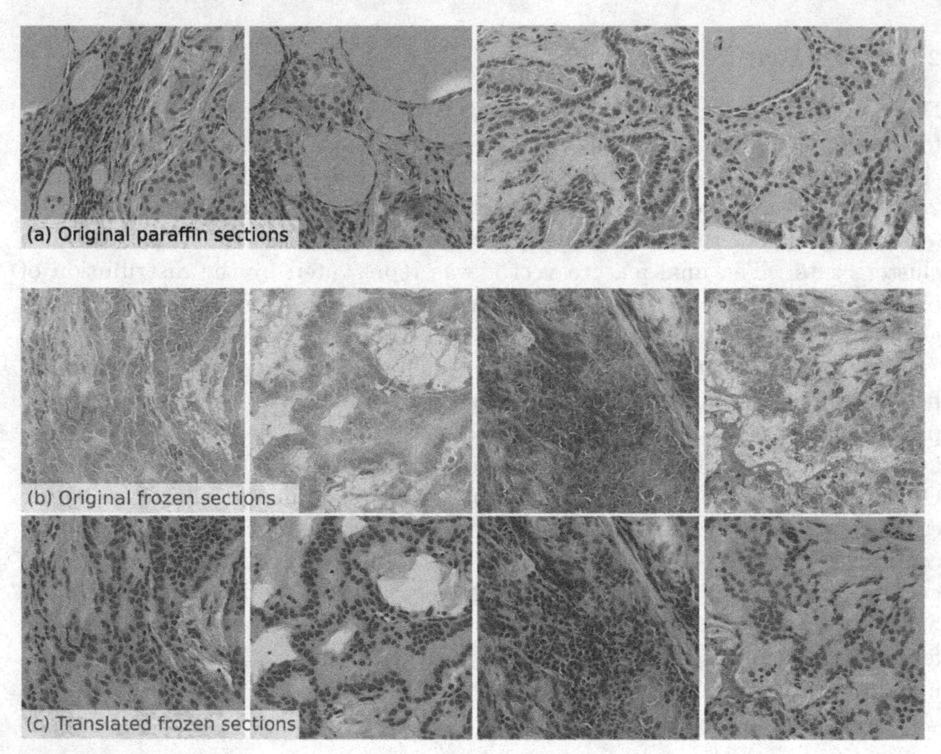

Fig. 3. Example patches with a size of 700 × 700 pixels showing (a) paraffin sections, (b) frozen sections and (c) the corresponding 'frozen-to-paraffin' sections (i.e. the original frozen sections were translated to the paraffin domain).

VS120-LD100 slide loader system. Overviews at a 2x magnification were generated to manually define scan areas, focus points were automatically defined and adapted if needed. Scans were performed with a 20x objective. Image files were stored in the Oympus vsi format based on lossless compression.

2.4 Experimental Details

The available data was randomly separated into training (80%) and test data (20%). The whole pipeline, including the separation was repeated 32 times, in order to achieve representative scores. Due to the almost balanced setting (see Sect. 2.3), the classification accuracy (mean and standard deviation) was finally reported.

Patch Extraction. Patches were randomly extracted from the whole slide image, based on uniform sampling. For each patch, we checked that at least 75% of the area was covered with tissue in order to exclude empty areas. We extracted a total of 512 patches with a size of 256 × 256 pixel per whole slide image.

Deep Feature Extraction. Due to the limited number of whole slide images, we did not train the feature extraction stage [7], but used a pre-trained network instead. Specifically, we use a ResNet18 pre-trained on the image-net challenge data, due to the high performance in previous work on similar data [3]. ResNet18 is particularly appropriate due to the rather low dimensional output of the convolutional layers (512 neurons). We directly forward the convolutional layers' output to the next stage.

Bag-of-Features Approach. The CNN features from the extracted patches of the training data were employed for training a k-means clustering model. The number of cluster centers was varied ($c \in \{16, 32, 64, 128\}$) to compare low and high dimensional descriptors and to study the effect in conjunction with data augmentation. As distance metric, the Euclidean distance was applied. Based on the cluster model, training and test feature vectors were clustered according to the closest distance to a cluster center and finally aggregated into one histogram for each whole slide image. The histograms constitute the final whole slide image-level feature vector for classification.

Classification Model. As classification model, we considered a linear SVM as well as an RBF kernel SVM. The cost factor c was optimized in inner cross validation ($c \in \{0.1, 1, 10, 100, 1000\}$). These rather basic models are used due to the small amount of training data needed. Two different approaches (a linear and a more flexible RBF approach) were here explicitly desired, in order to gain further insight.

3 Results

Quantitative results of the classification pipeline are shown in Fig. 4. For both classification models ((a, b) RBF kernel SVM and (c, d) linear SVM), three augmentation strategies (Aug1, Aug2, default (no augmentation)) and four different clustering strategies (C16, C32, C64, C128), as well as the mean accuracy including the standard deviations, are presented. While original paraffin sections exhibited the best scores in 32 out of 36 configurations, frozen-to-paraffin translation led to improvements, compared to original frozen sections, in 100% of configurations. The kernel SVM's scores were higher in 24 out of 36 configurations and also the maximum accuracy was higher (0.91 vs 0.87) with this classifier. Optimum clustering strategy varied, depending on the classifier and the imaging modality. Highest scores for paraffin (and overall) were achieved with C32 and C64. With frozen sections (and also frozen-to-paraffin translations), C128 exhibited the best performance. Data augmentation improved the accuracy in 14 out of 16 cases with paraffin sections, in 6 out of 16 cases with frozen section and in 12 out of 16 cases with translated sections.

The output of the image translation process is visualized for four example patches in Fig. 3. While the corresponding images ((b), (c)) show similar underlying tissue, the translated data (c) shows enhanced contrast of e.g. nuclei.

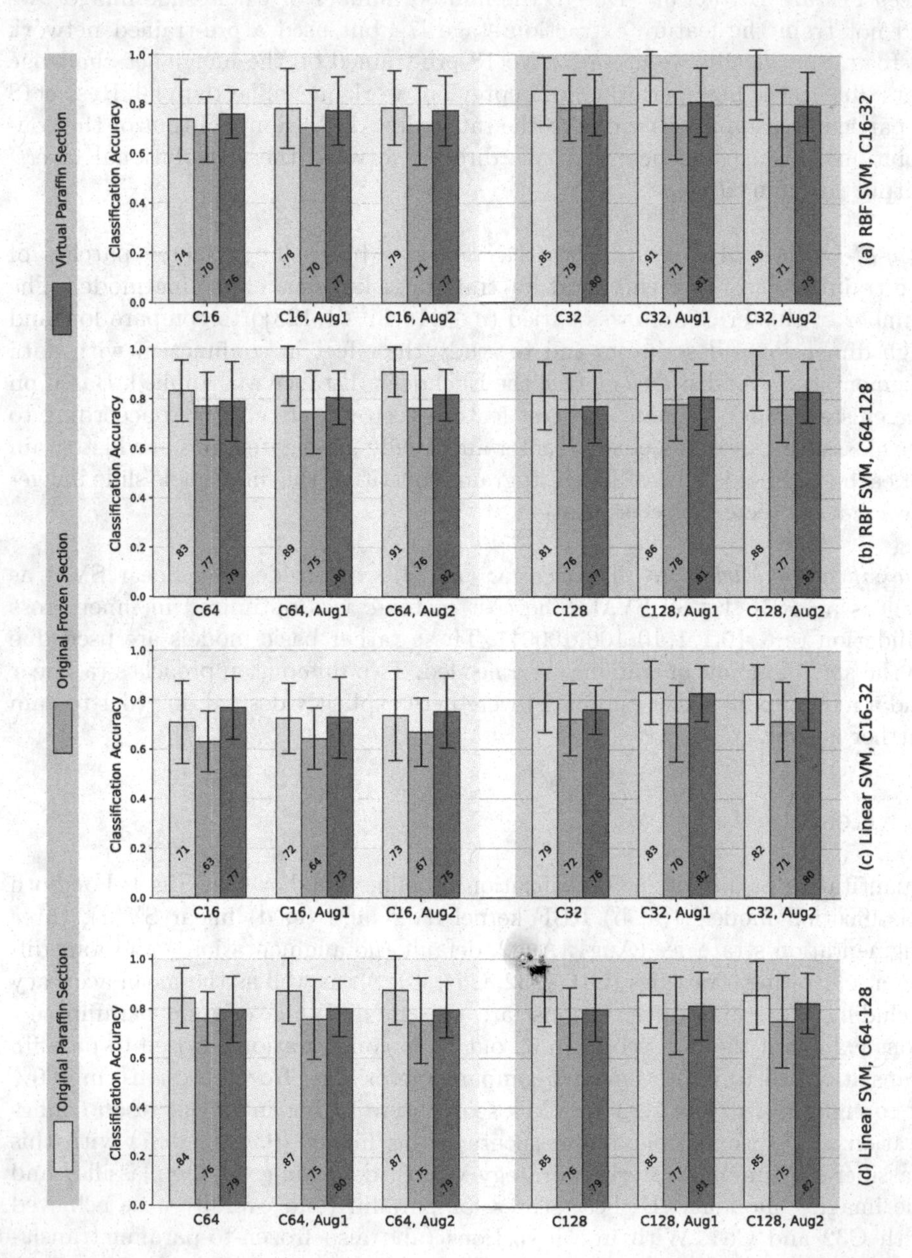

Fig. 4. Classification accuracy for different settings consisting of the number of histogram bins (C16, C32, C64, C128) and the data augmentation strategy (default no augmentation and the augmentation strategies Aug1 and Aug2) and the classification model.

4 Discussion

The first goal of this work was to study the effect of the underlying section type on the classification accuracy of thyroid cancer. We performed experiments with many different configurations, in order to be able to make general predictions and to be able to determine trends and correlations. Comparing overall classification accuracy of original paraffin and frozen section, we noticed that paraffin embedding exhibits the most appropriate section type for automated processing. Superior scores in case of paraffin sections are obtained completely independently of the underlying image analysis models' configuration. Based on these outcomes, we conclude that the lower visual quality of frozen section obviously not only affects manual assessment, but also computer-based decision support techniques. Accuracy of computer vision approaches, obviously similarly suffers from the less appropriate modality even though there is not drift between distribution considering training and test data. Potentially, if also the CNN feature extraction stage is adapted (and not only pre-trained on a different data set), this effect could be diminished. However, recent work indicates that also the performance of deep learning-based approaches depends on the quality of visual features [4]. As variability of frozen sections is typically higher, the effect is expected to be even stronger in case of multi-centered studies.

When translating the frozen section to paraffin, we noticed a clear improvement of accuracy. This effect was also prevalent, independent of the setting, but was particularly strong for low dimensional histograms (C16, C32) and also if a linear classification model was used. We expected that the image translation approach not only translates the data from one modality to the other, but also performed an effective kind of normalization (see also Fig. 3). This assumption was also confirmed by the fact that with specific settings (C16, linear SVM), the scores obtained with original paraffin sections were even outperformed. Geometric transforms caused by damaged (torn) tissue, could not be corrected.

The data augmentation strategy presented here improved scores in the majority of configurations with paraffin and frozen-to-paraffin sections. In the augmentation approach, there was a trade-off between diversity of data and inaccuracy in the distribution due to a reduced number of patches per histogram (in the bag-of-features approach). We assume that the negative performance of the data augmentation approach in case of original frozen sections is caused by the higher level of variability within the individual whole slide images. For both approaches, dealing with real or fake paraffin sections, one of the augmentation strategies exhibits the best configuration. Finally, there is also a trend that Aug2 with fewer samples per whole slide image and thereby higher variability in the training data, performs best with a higher number of cluster centers (e.g. C128). This is obviously due to the fact that the higher number of features require a higher variability in the training data.

5 Conclusion

In this work, we showed that the reduced image quality of frozen sections, which limits accuracy for pathologists, also leads to a clearly reduced classification accuracy of an automated classification system. This effect was observed widely independently of the underlying image analysis model. We further showed that image translation, from frozen to paraffin sections, based on a state-of-the-art generative adversarial network, was able to clearly increase the performance in case of frozen sections and reduced the gap to real paraffin sections. The proposed data augmentation strategy increased the scores of virtual paraffin sections in the classification setting even further. Finally, this work provides a strong motivation for performing a study with expert pathologists performing categorization of frozen sections and the corresponding translated sections, to investigate whether a similar effect is achieved in a clinical setting.

Acknowledgement. This work was partially funded by the County of Salzburg under grant number FHS-2019-10-KIAMed.

References

1. Almahairi, A., Rajeshwar, S., Sordoni, A., Bachman, P., Courville, A.C.: Augmented cycleGAN: learning many-to-many mappings from unpaired data. In: Proceedings of International Conference on Machine Learning (ICML 2018) (2018)
2. Bentaieb, A., Hamarneh, G.: Adversarial stain transfer for histopathology image analysis. IEEE Trans. Med. Imaging **37**(3), 792–802 (2018)
3. Dimitriou, N., Arandjelović, O., Caie, P.D.: Deep learning for whole slide image analysis: an overview. Front. Med. **6** (2019). https://doi.org/10.3389/fmed.2019.00264
4. Gadermayr, M., Gupta, L., Appel, V., Boor, P., Klinkhammer, B.M., Merhof, D.: Generative adversarial networks for facilitating stain-independent supervised and unsupervised segmentation: a study on kidney histology. IEEE Trans. Med. Imaging **38**(10), 2293–2302 (2019)
5. Gupta, L., Klinkhammer, B.M., Boor, P., Merhof, D., Gadermayr, M.: Stain independent segmentation of whole slide images: a case study in renal histology. In: Proceedings of the IEEE International Symposium on Biomedical Imaging (ISBI 2018) (2018)
6. Halicek, M., et al.: Head and neck cancer detection in digitized whole-slide histology using convolutional neural networks. Sci. Rep. **9**(1) (2019)
7. Hou, L., Samaras, D., Kurc, T.M., Gao, Y., Davis, J.E., Saltz, J.H.: Patch-based convolutional neural network for whole slide tissue image classification. In: Proceedings of the International Conference on Computer Vision (CVPR 2016) (2016)
8. Huang, X., Liu, M.Y., Belongie, S., Kautz, J.: Multimodal unsupervised image-to-image translation. In: Proceedings of the European Conference on Computer Vision (ECCV 2018) (2018)
9. Huber, G.F., et al.: Intraoperative frozen-section analysis for thyroid nodules. Arch. Otolaryngol.-Head Neck Surg. **133**(9), 874 (2007)
10. Isola, P., Zhu, J.Y., Zhou, T., Efros, A.A.: Image-to-image translation with conditional adversarial networks. In: Proceedings of the International Conference on Computer Vision and Pattern Recognition (CVPR 2017) (2017)

11. Leteurtre, E., et al.: Why do frozen sections have limited value in encapsulated or minimally invasive follicular carcinoma of the thyroid? Am. J. Clin. Pathol. **115**(3), 370–374 (2001)
12. Liu, M.Y., Breuel, T., Kautz, J.: Unsupervised image-to-image translation networks. In: Guyon, I., et al. (eds.) Advances in Neural Information Processing Systems (NIPS), pp. 700–708 (2017)
13. Najah, H., Tresallet, C.: Role of frozen section in the surgical management of indeterminate thyroid nodules. Gland Surg. **8**(S2), 112–117 (2019)
14. Osamura, R.Y., Hunt, J.L.: Current practices in performing frozen sections for thyroid and parathyroid pathology. Virchows Arch. **453**(5), 433–440 (2008)
15. Park, T., Efros, A.A., Zhang, R., Zhu, J.Y.: Contrastive learning for conditional image synthesis. In: Proceedings of the European Conference on Computer Vision (ECCV 2020) (2020)
16. Udelsman, R., Westra, W.H., Donovan, P.I., Sohn, T.A., Cameron, J.L.: Randomized prospective evaluation of frozen-section analysis for follicular neoplasms of the thyroid. Ann. Surg. **233**(5), 716–722 (2001)
17. Wang, S., Yang, D.M., Rong, R., Zhan, X., Xiao, G.: Pathology image analysis using segmentation deep learning algorithms. Am. J. Pathol. **189**(9), 1686–1698 (2019)
18. Wolterink, J.M., Dinkla, A.M., Savenije, M.H.F., Seevinck, P.R., van den Berg, C.A.T., Išgum, I.: Deep MR to CT synthesis using unpaired data. In: Tsaftaris, S.A., Gooya, A., Frangi, A.F., Prince, J.L. (eds.) SASHIMI 2017. LNCS, vol. 10557, pp. 14–23. Springer, Cham (2017). https://doi.org/10.1007/978-3-319-68127-6_2
19. Yi, Z., Zhang, H., Tan, P., Gong, M.: DualGAN: unsupervised dual learning for image-to-image translation. In: Proceedings of the IEEE International Conference on Computer Vision (ICCV 2017) (2017)
20. Zhu, J.Y., Park, T., Isola, P., Efros, A.A.: Unpaired image-to-image translation using cycle-consistent adversarial networks. In: Proceedings of the International Conference on Computer Vision (ICCV 2017) (2017)

SequenceGAN: Generating Fundus Fluorescence Angiography Sequences from Structure Fundus Image

Wanyue Li[1,2], Yi He[1,2], Wen Kong[1,2], Jing Wang[1,2], Guohua Deng[3], Yiwei Chen[1], and Guohua Shi[1,2,4(✉)]

[1] Suzhou Institute of Biomedical Engineering and Technology, Chinese Academy of Sciences, Suzhou, China

[2] School of Biomedical Engineering (Suzhou), Division of Life Sciences and Medicine, University of Science and Technology of China, Hefei, China

[3] Department of Ophthalmology, The Third People's Hospital of Changzhou, Changzhou, China

[4] Center for Excellence in Brain Science and Intelligence Technology, Chinese Academy of Sciences, Shanghai, China

Abstract. Fundus fluorescein angiography (FA) is an indispensable procedure that can investigate the integrity of retina vasculature. Fluorescein angiograms progress through five phases: pre-arterial, arterial, arteriovenous, venous, and late, and each phase could be an important diagnostic basis for retina-related disease. However, the FA imaging technique may provide risks of harm to the patients. To help physicians reduce the potential risks of diagnosis, we proposed "SequenceGAN", a novel sequential generative adversarial network that aims to generate FA sequences of critical phases from a structure fundus image. Moreover, a feature-space loss is applied to ensure the generated FA sequences with a better visual effect. The proposed method was qualitatively and quantitatively compared with existing FA image generation methods and image translation methods. The experimental results indicate that the proposed model has better performance on the generation of retina vascular, leakage structures, and characteristics of each angiogram phase, and thus indicates potential value for application in clinical diagnosis.

Keywords: Fundus fluorescein angiography · Sequential generative adversarial network · Structure fundus image · Feature-space loss

1 Introduction

Fundus fluorescein angiography (FA) can reflect the damaged state of the retinal barrier in the living human eyes and dynamically capture the physiological and pathological conditions from the large vessels of the retina to capillaries [1], which is regarded as the "gold standard" of retinal disease diagnosis [2]. Fluorescein angiograms progress through five phases: pre-arterial, arterial, arteriovenous, venous, and late [3]. In the arterial and venous phase of angiograms, whether the filling time of retinal vessels is

D. Svoboda et al. (Eds.): SASHIMI 2021, LNCS 12965, pp. 110–120, 2021.
https://doi.org/10.1007/978-3-030-87592-3_11

normal or not is an important diagnostic basis for retinal vascular occlusion diseases (such as central retinal artery occlusion [4] and retinal vein occlusion [5]). And the late phase of FA is crucial for the diagnosis of most retina-related diseases. Intravenous injection of sodium fluorescein contrast agent is an essential procedure for FA imaging. However, sodium fluorescein contrast agents are not suitable for all patients, it is banned for severe hypertension, heart disease, and other patients, and it also has potential adverse effects [6, 7]. Therefore, generating the critical phase FA sequences from the fundus structure image could be a good idea to help physicians diagnose with smaller potential risks in patients.

Generating FA sequences from a fundus structure image can be defined as an image translation task. With the development of deep learning and generative adversarial networks (GANs), image translation technique has also been greatly developed. Algorithms that tackle image translation can be divided into one-to-one and one-to-many translation methods. One-to-one image translation refers to translating images from one domain to another, and the most widely known methods include pix2pix [8] and CycleGAN [9]. Where, pix2pix is a paired method that requires paired data for image translation, and CycleGAN is an unpaired method. One-to-many image translation methods translate one domain image to multi-domain or multi-modal, which is a hot research area in recent years. Zhu et al. [10] proposed BicycleGAN to translate one domain image to another with multimodal using paired data. Choi et al. [11] proposed a model named "StarGAN" that can perform unsupervised image-to-image translation for multiple domains using only one generator and discriminator. As for FA image, Schiffers et al. [12] proposed a CycleGAN based method to generate FA image from retinal fundus image. Hervella et al. [13] utilized a generator composed of an encoder, 9 residual blocks, and a decoder to synthesize FA image based on paired and unpaired data, respectively. Li et al. [1] proposed a conditional GAN-based FA image generation method with a novel saliency loss. All of the existing methods [1, 12–16] mainly focus on the generation of single frame FA image of single phase. However, for retinal vascular occlusion diseases, it is not enough for physicians to only focus on a single frame FA image of the single phase, and the angiographic performance of the multiple critical phases should be considered comprehensively.

In this paper, we proposed a novel framework, called "SequenceGAN", to generate FA sequences of critical phases from an input fundus structure image. To ensure the better visual effect of generated sequences, a feature-space loss is applied. Since several studies [1, 13] have demonstrated that the unpaired methods have not yet been able to accurately generate retinal vascular and pathological structures, the proposed method is a paired method. The contributions of this work can be summarized as follows:

- As far as our knowledge goes, it is the first time to present a study about generating FA sequences of critical phases from a single structure fundus image.

- We proposed SequenceGAN, a novel sequential generative adversarial network that learns the mappings between one domain image and another domain image sequences.
- We provide both qualitative and quantitative results on the proposed SequenceGAN, the existing FA image generation methods, and the state-of-the-art image translation methods, which shows the superiority of the proposed method.

2 Method

2.1 Datasets

The datasets used in this study were taken with Spectralis HRA equipment (Heidelberg Engineering, Heidelberg, Germany) between August 2011 and June 2019 at the Third People's Hospital of Changzhou (Jiangsu, China), and we call this dataset "HRA sequence dataset". Since the arterial, venous, and late phase FA images are more helpful for most retina-related diseases diagnosis, we select the FA images captured from the arterial phase (11 to 15 s), venous phase (16 to 20 s), and late phase (5 to 6 min) as the label FA sequences and one image per phase. This dataset contained 126 groups of the image from 126 eyes of 108 patients (46 female, 62 male, ranging in age from 17 to 72 years), each group contained a structure fundus image and three corresponding FA images from three phases, of which 26 groups were randomly selected to comprise the testing set, and the remainder 100 groups were employed as the training set. The field of views of these images including 30°, 45°, 60°, and the resolution of each image is 768 × 768 pixels.

The proposed method is a paired method that requires paired data. However, the collected structure fundus image and its corresponding FA image sequences always unaligned, thus a dataset preprocessing approach proposed by Li et al. [1] is applied. After dataset processing, we finally have 2,245 groups of 512 × 512 pixels paired image patches used for training, and 674 groups for testing.

2.2 SequenceGAN for FA Sequences Generation

Our goal is to train a network that can accurately generate FA sequences of critical phases (arterial, venous, and late phase) from a single input structure fundus image. To achieve this, we designed a sequential generative adversarial network that has three generators and discriminators, as depicted in Fig. 1. G_O is trained to translate a random noise n into an output arterial phase FA image with the input structure fundus image as condition, $G_O(I_S, n) \rightarrow I_{FO}$. G_I is designed to generate a venous phase FA image from a generated arterial phase FA image and an input structure fundus image, $G_I(I_S, G_O(I_S, n)) \rightarrow I_{FI}$. And G_2 is to generate a late phase FA image from a synthetic venous phase FA image and a structure fundus image, $G_2(I_S, G_I(I_S, G_O(I_S, n))) \rightarrow I_{F2}$. The input structure fundus image is also the conditional image for both G_I and G_2. Three discriminators D_O, D_I, and D_2 are used to distinguish between real and generated FA images with the condition of structure fundus image. The configuration of each sub-model is similar to the model proposed in work [1], which uses 9 residual blocks as the core of generators, and 70 × 70 PatchGANs as the discriminators. Noted that the generators and discriminators in the proposed model share the same architecture, and all the GANs are trained simultaneously.

To ensure that the generated FA image resembles the label image, three loss functions are applied for each sub-GAN, which is the adversarial loss L_{GAN}, pixel-space loss L_{pixel}, and feature-space loss L_{percep}, and the full loss function can be written as follows:

$$L = \alpha L_{GAN} + \beta L_{pixel} + \gamma L_{percep}, \tag{1}$$

where α, β, and γ are the experimentally determined hyperparameters that control the effect of adversarial loss, pixel-space loss, and feature-space loss. Pursuing balance among these losses is not a trivial task. The value of α and β were preliminarily determined according to works [1, 8]. As for the parameter of feature-space loss, we conducted multiple experiments to find the most suitable value to achieve the best performance; and experimental analysis will be illustrated in Sect. 3.3. Fully considering the qualitative and quantitative results, we set $\alpha = 1, \beta = 100, \gamma = 0.001$ in this task.

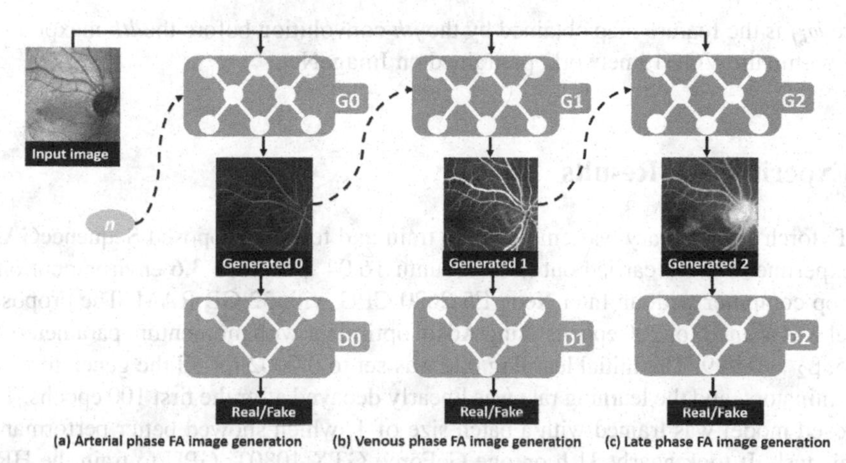

Fig. 1. The architecture of SequenceGAN. Three generators and discriminators are designed to generate three FA images of three phases. (a) Arterial phase FA image generation; (b) Venous phase FA image generation; (c) Late phase FA image generation. (*The proposed model is based on conditional GAN and structure fundus image is the condition of each sub-model.)

Adversarial Loss: Since we have three generators and discriminators, the full GAN loss can be defined as follows:

$$\begin{aligned} L_{GAN} = & \mathop{\mathrm{E}}_{I_s, I_{F0}}[log D_0(I_S, I_{F0})] + \mathop{\mathrm{E}}_{I_S, n}[log(1 - D_0(I_S, G_0(I_S, n)))] \\ & + \mathop{\mathrm{E}}_{I_s, I_{F1}}[log D_1(I_S, I_{F1})] + \mathop{\mathrm{E}}_{I_S, n}[log(1 - D_1(I_S, G_1(I_S, G_0(I_S, n))))] \\ & + \mathop{\mathrm{E}}_{I_s, I_{F2}}[log D_2(I_S, I_{F2})] + \mathop{\mathrm{E}}_{I_S, n}[log(1 - D_2(I_S, G_2(I_S, G_1(I_S, G_0(I_S, n)))))]. \end{aligned} \tag{2}$$

Pixel-Space Loss: Similarly to work [1], an L1 loss is adopted to encourage the generated image to be close to the target image, the full pixel-space loss is described as:

$$L_{pixel} = \underset{I_S,n,I_{F0}}{E}[\|G_0(I_S,n) - I_{F0}\|_1] + \underset{I_S,n,I_{F1}}{E}[\|G_1(I_S,G_0(I_S,n)) - I_{F1}\|_1] + \underset{I_S,n,I_{F2}}{E}[\|G_2(I_S,G_1(I_S,G_0(I_S,n))) - I_{F2}\|_1]. \quad (3)$$

Feature-Space Loss: Perceptual loss [17] is used as the feature-space loss, which calculates the difference between the CNN features of the generated image and the target image to help reconstruct textural and detailed information. It can be written as:

$$L_{percep} = \underset{I_S,n,I_{F0}}{E}\left[\|\varphi_{i,j}(G_0(I_S,n)) - \varphi_{i,j}(I_{F0})\|_2^2\right] + \underset{I_S,n,I_{F1}}{E}\left[\|\varphi_{i,j}(G_1(I_S,G_0(I_S,n))) - \varphi_{i,j}(I_{F1})\|_2^2\right] + \underset{I_S,n,I_{F2}}{E}\left[\|\varphi_{i,j}(G_2(I_S,G_1(I_S,G_0(I_S,n)))) - \varphi_{i,j}(I_{F2})\|_2^2\right], \quad (4)$$

where $\varphi_{i,j}$ is the feature map obtained by the *jth* convolution before the *ith* maxpooling layer within the VGG19 network, pretrained on ImageNet.

3 Experimental Results

The Pytorch 1.7.0 library was employed to train and test the proposed SequenceGAN. All experiments were carried out in an Ubuntu 16.04 + Python 3.6 environment on a desktop computer with an Intel Xeon E5-2620 CPU with 32 GB RAM. The proposed model was trained for 200 epochs using Adam optimizer with momentum parameters $\beta_1 = 0.5$, $\beta_2 = 0.999$. The initial learning rate was set to 0.0002 for all the generators and discriminators, and the learning rate was linearly decayed after the first 100 epochs. The proposed model was trained with a batch size of 1, which showed better performance on this task. It took nearly 31 h on one GeForce GTX 1080Ti GPU to train the HRA sequence dataset.

3.1 Evaluation Metrics

The testing results were quantitatively evaluated with the criteria of peak signal-to-noise ratio (PSNR), structural similarity (SSIM), and perceptual distance[18]. PSNR and SSIM were commonly used in FA image generation studies [1, 13, 14]. These two metrics are defined as follows:

$$MSE = \sum_{n=1}^{N}(x^n - y^n)\Big/N \quad (5)$$

$$PSNR = 10 \times log_{10}\left(255^2\Big/MSE\right) \quad (6)$$

$$SSIM(x,y) = \frac{(2\mu_x\mu_y + c_1)(2\sigma_{xy} + c_2)}{(\mu_x^2 + \mu_y^2 + c_1)(\sigma_x^2 + \sigma_y^2 + c_2)}, \quad (7)$$

where N is the size of image; x^n and y^n are the nth pixels of the generated image x and the target image y; μ_x and μ_y are the means of x and y; σ_x^2 and σ_y^2 are the variance of x and y; σ_{xy} is the covariance of x and y.

Perceptual distance is used for measuring the perceptual quality of the generated images, and partially coincides with the human perceptron. A smaller perceptual distance implies better perceptual quality and visual effects. Given a patch of generated image x and target image y, the perceptual distance can be calculated as:

$$d(x, y) = \sum_l \frac{1}{H_l W_l} \sum_{h,w} \left\| w_l \odot (\hat{x}_{hw}^l - \hat{y}_{hw}^l) \right\|_2^2, \tag{8}$$

where $\hat{x}_{hw}^l$ and $\hat{y}_{hw}^l$ are the feature stacks extracted from layer l of a specific network and unit-normalized in the channel dimension; w_l is the scale vector; H_l and W_l are the dimensions of image patches.

3.2 Results Comparison on FA Sequences Generation

To demonstrate the effectiveness of the proposed model, we compared the proposed method with those of two single frame FA image generation methods [1, 13] and three state-of-the-art image translation methods (pix2pix [8], BicycleGAN [10], and StarGAN [11]). Where pix2pix [8] is a one-to-one image translation method and has been widely used in medical image translation tasks. BicyleGAN [10] and StarGAN [11] are two one-to-many methods, of which StarGAN is an unsupervised method that can translate multi-domain images without the use of paired data. For fair comparison, these five methods were all trained and tested on our dataset.

Figure 2 illustrates the qualitative comparison results with state-of-the-art methods. The FA images in the first and fourth row, the second and fifth row, and the third and sixth row are the generated images of arterial, venous, and late phase. For better comparison, we zoom in on the local regions with red boxes. As shown in Fig. 2(g), the images generated by StarGAN only roughly reflect the appearance of the FA images, and cannot accurately generate the pathological features of images. As described in Fig. 2 (b)–(f), these five methods that need paired data can generate the details of retinal vascular and even the retinal vessels that are not filling in the arterial phase. Furthermore, the retina vascular generated by the proposed model (Fig. 2(b)) are more accurate and more similar to real FA images. Moreover, the proposed SequenceGAN performs better on the generation of leakage structures when compared with pix2pix, BicycleGAN, and the existing FA image generation methods [1, 13], as depicted in the third row of Fig. 2(b)–(f).

For quantitative comparison, the FA images generated by SequenceGAN, pix2pix, Li's method, Hervella's method, BicycleGAN, and StarGAN were all evaluated with the metrics of PSNR, SSIM, and perceptual distance. As seen in Table 1, the proposed SequenceGAN outperforms other five methods. The average PSNR of the proposed method is 0.08 dB, 0.01 dB, 0.13 dB, 1.92 dB, and 3.13 dB higher than those of pix2pix, Li's method [1], Hervella's method [13], BicycleGAN, and StarGAN, respectively; and the average SSIM is 0.0096, 0.0155, 0.0113, 0.0413, and 0.1181 higher, respectively. The average perceptual distance of the proposed method is 0.0137, 0.0066, 0.0106, 0.0257, and 0.1407 lower than those of pix2pix, Li's method [1], Hervella's method

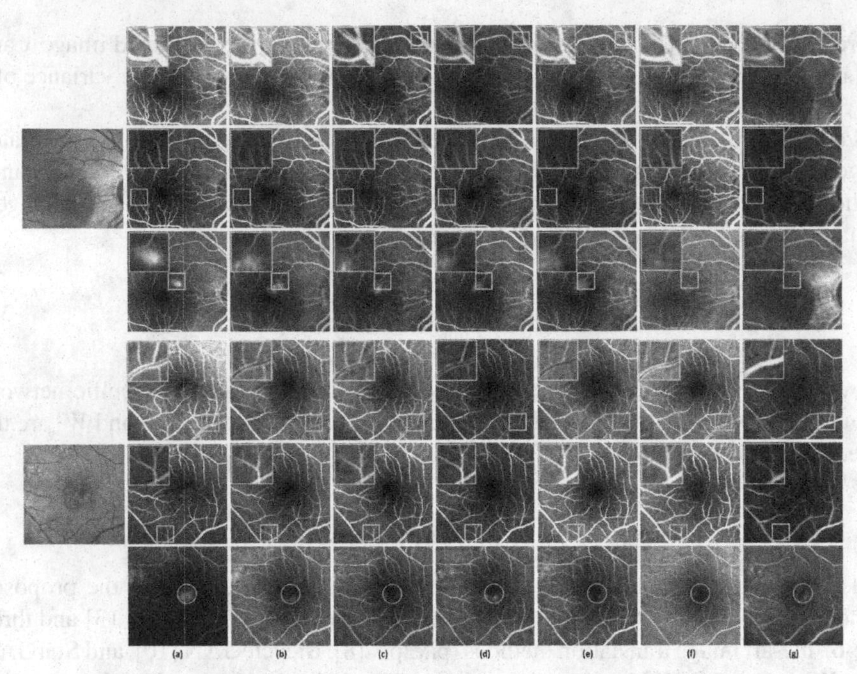

Fig. 2. Qualitative comparison with state-of-the-art methods. The FA images in the first and fourth row, the second and fifth row, and the third and sixth row are the generated images of arterial, venous, and late phase, respectively. (a) Real FA image. FA images generated from: (b) the proposed SequenceGAN; (c) pix2ix; (d) Li's method [1]; (e) Hervella's method [13]; (f) BicycleGAN; and (g) StarGAN. (Zooming-in the figure will provide a better look at the image details).

[13], BicycleGAN, and StarGAN, respectively. This indicates that the proposed method can generate FA images with better image quality and visual effect.

Table 1. Performance comparison with state-of-the-art methods. (*Data are expressed as mean ± standard deviation)

	PSNR(dB)	SSIM	Perceptual distance
Sequence GAN	**15.63 ± 2.43**	**0.7194 ± 0.0930**	**0.3278 ± 0.0393**
pix2pix	15.55 ± 2.41	0.7098 ± 0.0883	0.3415 ± 0.0440
Li's method [1]	15.62 ± 3.00	0.7039 ± 0.1013	0.3344 ± 0.0439
Hervella's method [13]	15.50 ± 2.42	0.7081 ± 0.0873	0.3384 ± 0.0405
BicycleGAN	13.71 ± 3.31	0.6781 ± 0.1022	0.3535 ± 0.0438
StarGAN	12.50 ± 2.52	0.6013 ± 0.1051	0.4685 ± 0.0410

3.3 Ablation Study

The baseline of the proposed SequenceGAN is the model without the FA image generated from former phase as input (only the random noise n as input like Fig. 1(a)) and the perceptual loss. Thus, we conducted an ablation study to validate the inclusion of input FA image and perceptual loss. Furthermore, we also did multiple experiments to analyze how the regularization parameter γ for the perceptual loss term influences the results of the proposed model.

To discuss the respective role of input FA image and perceptual loss, both qualitative and quantitative comparisons are provided. As shown in Table 2, although the PSNR and SSIM of the model with input FA image and perceptual loss are lower than that without perceptual loss, this model has lower perceptual distance and performs best on the generation of retinal vessels and leakage structures (Fig. 3(f)). This implies that perceptual loss can help generate the details of the image and obtain a better visual effect. Moreover, using the FA image generated from former phase as input can also help the generation of important information (Fig. 3(f)) and get higher PSNR and SSIM.

To determine the most suitable parameter of perceptual loss, we conducted multiple experiments and settle on a suitable range of γ that ranges from 0 to 0.01. Figure 4 illustrates the quantitative results of the proposed model with different perceptual loss parameters. The top-4 best results ($\gamma = 0, 0.001, 0.007$, and 0.009) are highlighted in red font, and the qualitative comparisons are described in Fig. 3. It can be seen in Fig. 3(a)–(b) and 3(d)–(f), the proposed model with $\gamma = 0.001$ can obtain the best performance on the generation of retinal vascular and leakage structures and has a better visual effect.

Table 2. Ablation study performance of the proposed SequenceGAN (*Data are expressed as mean ± standard deviation)

baseline	Input FA image	Perceptual loss	PSNR(dB)	SSIM	Perceptual distance
√			15.13 ± 3.48	0.7067 ± 0.1175	0.3419 ± 0.0404
√	√		**15.89 ± 3.19**	**0.7289 ± 0.1013**	0.3309 ± 0.0441
√	√	√	15.63 ± 2.43	0.7194 ± 0.0930	**0.3278 ± 0.0393**

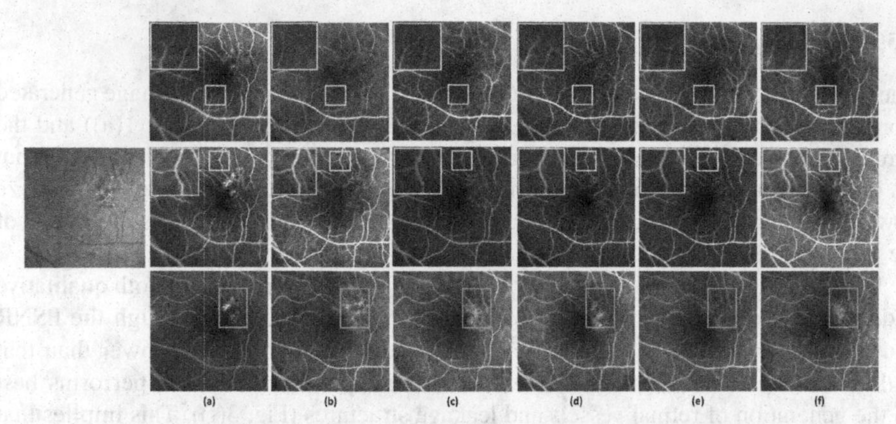

Fig. 3. Results of ablation study. The FA images in the first to the third rows are the generated images of arterial, venous, and late phase. (a) Real FA images; (b) FA images generated by baseline. FA images generated from the proposed model: (c) without perceptual loss ($\gamma = 0$); (d) with $\gamma = 0.009$; (e) with $\gamma = 0.007$; and (f) with $\gamma = 0.001$. (Zooming-in the figure will provide a better look at the image details).

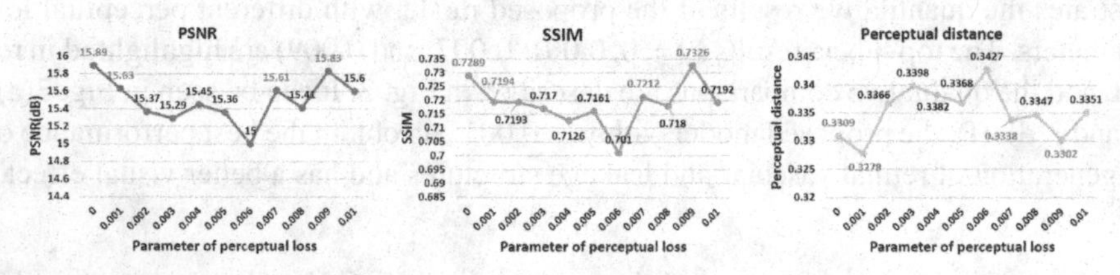

Fig. 4. Statistic diagram of the results produced with different parameter of perceptual loss. (*the top-4 best results are highlighted in red font. And zooming-in the figure will provide a better look at the specific value). (Color figure online)

4 Discussion and Conclusion

Fundus fluorescein angiography (FA) can dynamically capture the physiological and pathological conditions of the retina, and the FA images of each phase could serve as an important basis for clinical diagnosis. However, the contraindication and potential adverse effects of FA imaging limit its usage to some extent. With the development of image translation techniques, an idea about generating FA image from structure fundus image to help make up the defects of FA imaging has been proposed [1, 12–16]. However, the existing methods mainly focus on the generation of a single frame FA image, which is not sufficient for the diagnosis of some retinal diseases. In this work, we proposed a novel model, named "SequenceGAN", to generate FA sequences from a single structure fundus image.

In this work, we choose the FA images in arterial, venous, and late phase as the target images, which are crucial phases for most retina disease diagnosis. As shown in Fig. 2, the proposed model has a good overall performance on the generation of characteristics of each angiogram phase, however, it performs unsatisfied on leakage details (the last row of Fig. 2(b)). Moreover, the proposed SequenceGAN requires three generators and discriminators to generate three FA sequences, i.e., it will require more generators and discriminators to generate more sequences, which is time-consuming and resource intensive. Therefore, designing a more effective and efficient model is the main task in the following study. In addition, limited by the dataset, this work only preliminarily achieves the method of FA sequences generation. At present, it is not possible to make quantitative evaluation of the generated images from the perspective of medical diagnosis, which needs to be further studied in the follow-up work.

To conclude, the proposed SequenceGAN model has better overall performance on the generation of retina vessel details, leakage structures, and the characteristics of each angiogram phase when compared with other methods, and thus shows potential significance for application in clinical diagnosis.

Acknowledgement. This work is supported by the National Key Research and Development Program of China (2016YFF0102002) and the National Natural Science Foundation of China (61605210, 62075235).

References

1. Li, W.: Generating fundus fluorescence angiography images from structure fundus images using generative adversarial networks. Proc. Mach. Learn. Res. **121**, 1–16 (2020)
2. O'Toole, L.: Fluorescein and ICG angiograms: still a gold standard. Acta Ophthalmol. Scand. **85** (2007)
3. Dolan, B.J.: Fluorescein and ICG angiography textbook and atlas. Optom. Vis. Sci. **76**, 520 (1999)
4. Varma, D.D., Cugati, S., Lee, A.W., Chen, C.S.: A review of central retinal artery occlusion: clinical presentation and management. Eye **27**, 688–697 (2013)
5. Wong, T.Y., Scott, I.U.: Retinal-vein occlusion. N. Engl. J. Med. **363**, 2135–2144 (2010)
6. Yannuzzi, L.A., et al.: Fluorescein angiography complication survey. Ophthalmology **93**, 611–617 (1986)
7. Musa, F., Muen, W.J., Hancock, R.: Adverse effects of fluorescein angiography in hypertensive and elderly patients. Acta Ophthalmol. Scand. **84**, 740–742 (2006)
8. Isola, P.: Image-to-image translation with conditional adversarial networks. In: Proceedings of the IEEE Conference on Computer Vision and Pattern Recognition. CVPR (2017)
9. Zhu, J.Y., Park, T., Isola, P., Efros, A.A.: Unpaired image-to-image translation using cycle-consistent adversarial networks. In: Proceedings of the IEEE International Conference on Computer Vision (2017)
10. Zhu, J.: Toward multimodal image-to-image translation. In: Proceedings of the 31st International Conference on Neural Information Processing Systems, pp.465–476 (2017)
11. Choi, Y., Choi, M., Kim, M., Ha, J.W., Kim, S., Choo, J.: StarGAN: unified generative adversarial networks for multi-domain image-to-image translation. In: Proceedings of the IEEE Computer Society Conference on Computer Vision and Pattern Recognition (2018)
12. Schiffers, F., Yu, Z., Arguin, S., Maier, A., Ren, Q.: Synthetic fundus fluorescein angiography using deep neural networks. In: Maier, A., Deserno, T., Handels, H., Maier-Hein, K., Palm, C., Tolxdorff, T. (eds.) Bildverarbeitung für die Medizin 2018. Informatik aktuell, pp. 234–238. Springer, Heidelberg (2018). https://doi.org/10.1007/978-3-662-56537-7_64
13. Hervella, Á.S.: Deep multimodal reconstruction of retinal images using paired or unpaired data. In: International Joint Conference on Neural Networks (IJCNN), pp. 1–8. IEEE (2019)

14. Li, K., Yu, L., Wang, S., Heng, P.-A.: Unsupervised retina image synthesis via disentangled representation learning. In: Burgos, N., Gooya, A., Svoboda, D. (eds.) SASHIMI 2019. LNCS, vol. 11827, pp. 32–41. Springer, Cham (2019). https://doi.org/10.1007/978-3-030-32778-1_4
15. Hervella, Á.S., Rouco, J., Novo, J., Ortega, M.: Retinal image understanding emerges from self-supervised multimodal reconstruction. In: Frangi, A.F., Schnabel, J.A., Davatzikos, C., Alberola-López, C., Fichtinger, G. (eds.) MICCAI 2018. LNCS, vol. 11070, pp. 321–328. Springer, Cham (2018). https://doi.org/10.1007/978-3-030-00928-1_37
16. Kamran, S.A., Fariha Hossain, K., Tavakkoli, A., Zuckerbrod, S., Baker, S.A., Sanders, K.M.: Fundus2Angio: a conditional GAN architecture for generating fluorescein angiography images from retinal fundus photography. In: Bebis, G., et al. (eds.) ISVC 2020. LNCS, vol. 12510, pp. 125–138. Springer, Cham (2020). https://doi.org/10.1007/978-3-030-64559-5_10
17. Johnson, J., Alahi, A., Fei-Fei, L.: Perceptual losses for real-time style transfer and super-resolution. In: Leibe, B., Matas, J., Sebe, N., Welling, M. (eds.) ECCV 2016. LNCS, vol. 9906, pp. 694–711. Springer, Cham (2016). https://doi.org/10.1007/978-3-319-46475-6_43
18. Zhang, R., Isola, P., Efros, A.A., Shechtman, E., Wang, O.: The unreasonable effectiveness of deep features as a perceptual metric. In: Proceedings of the IEEE Conference on Computer Vision and Pattern Recognition, pp. 586–595 (2018)

Cerebral Blood Volume Prediction Based on Multi-modality Magnetic Resonance Imaging

Yongsheng Pan[1,2], Jingyu Huang[1], Bao Wang[3], Peng Zhao[4], Yingchao Liu[4], and Yong Xia[1,2](✉)

[1] School of Computer Science and Engineering, Northwestern Polytechnical University, Xi'an 710072, China
yxia@nwpu.edu.cn
[2] Research and Development Institute of Northwestern, Polytechnical University in Shenzhen, Shenzhen 518057, China
[3] Department of Radiology, Qilu Hospital of Shandong University, Jinan 250012, China
[4] Department of Neurosurgery, Shandong Provincial Hospital Afiliated to Shandong First Medical University, Jinan 250021, China

Abstract. Cerebral blood volume (CBV) refers to the blood volume of a certain brain tissue per unit time, which is the most useful parameter to evaluate intracranial mass lesions. However, the current CBV measurement methods rely on blood perfusion imaging technology which has obvious shortcomings, i.e., long imaging time, high cost, and great discomfort to the patients. To address this, we attempt to utilize some techniques to synthesize the CBV maps from multiple MRI sequences, which is the least harmful imaging technology currently, so as to reduce the time and cost of clinical diagnosis as well as the patients' discomfort. Two image synthesis techniques are investigated to synthesize the CBV maps on our collection of 103 groups of multiple MRI modalities of 70 subjects. The experimental results on various modality combinations demonstrate that our redesigned algorithms are possible to synthesize promising CBV maps, which is a good start of developing efficient and cheaper CBV prediction system.

Keywords: Cerebral blood volume · Medical image synthesis · Generative adversarial network

1 Introduction

Cerebral blood volume (CBV) [5], which is related to tumor microvessel density (MVD) [7], is the most useful parameter to evaluate intracranial mass lesions [13]. It refers to the blood volume of a certain brain tissue per unit time, which is calculated based on the enclosed area under the time-density curve. Tumors

Y. Pan and J. Huang contribute equally.

D. Svoboda et al. (Eds.): SASHIMI 2021, LNCS 12965, pp. 121–130, 2021.
https://doi.org/10.1007/978-3-030-87592-3_12

undergoing fast neovascularization to support the active proliferation of tumor cells, is the basis of tumor CBV increase [3]. Accordingly, CBV measurement of brain tumors is helpful for evaluating the degree of neovascularization with malignancy and grade, identifying tumor-like lesions and monitoring treatment effects, etc. The morphology of blood vessels and the degree of neovascularization are important basis for distinguishing the types of intracranial tumors and determining the degree of invasion. The CBV chart of living body, which reflects the degree of vascularization can be used to indirectly judge tumor neovascularization.

The current CBV measurement methods rely on blood perfusion weighted imaging (PWI) technology [10]. For example, dynamic magnetic sensitive contrast enhancement [9] requires intravenous injection of an exogenous tracer, and rapid imaging is used to obtain a series of images of the contrast agent passing through the examined tissue for the first time. Actually, these steps are very cumbersome and have obvious shortcomings, e.g., long imaging time, high cost, and great discomfort to the patient [8]. If the corresponding CBV image can be directly obtained through some simple means, the burden on the operator can be greatly reduced. At the same time, if the use of contrast agent can be avoided, the patient's discomfort can be greatly reduced and medical expenses can be saved. Therefore, the study of non-invasive methods has important clinical significance.

Recent years, image synthesis techniques have made great progress in medical imaging, which could synthesize images among various modalities. Therefore, it is possible to synthesize CBV images from other non-invasive images. Currently, Magnetic Resonance (MR) is the least harmful imaging technology, where multiple MR modalities (e.g., T1 weighted imaging (T1WI), T2 weighted imaging (T2WI), Contrast enhanced T1WI (T1_C), T2-weighted-fluid-attenuated Inversion recovery (T2_F), apparent diffusion coefficient (ADC)) can reflect various aspects of information [1]. These multi-modality images may contain shared information with CBV images, thus may be used to synthesize CBV images. Accordingly, we attempt to synthesis the CBV images from multiple-modality MR images, thus develop non-invasive and efficient CBV measurement.

In this paper, we propose to use the generative adversarial network(GAN)-based image synthesis techniques to synthesize the CBV maps from multiple-modality MR images. To address this, we collected 103 groups of multiple MRI modalities of 70 subjects, and reimplement the sense-consistency generative adversarial network (SGAN) [12] and the 3D encoder-decoder with only mean absolute error loss (MAE-ED) [2] to adopt them to our dataset. Our experiments on various combinations of different modalities demonstrate that it is possible to synthesize CBV maps from multiple-modality MR images, which is potentially useful in reducing the patients' discomfort as well as the diagnosis cost.

2 Materials and Methods

2.1 Participants

A total of 85 consecutive patients with brain metastases, all of whom had a pathologically confirmed diagnosis of lung cancer, were retrospectively identified from January 2017 to May 2019 at the Gamma Knife Radiosurgery Centre. All patients undergo unified multimodality MRI examinations. In addition, part of them undergo follow-up MRI examinations. In all, 15 patients were excluded for the following reasons: (1) failed or unqualified CBV map generation (6 patients); and (2) artifacts on MRI (9 patients). Finally, 70 patients (mean age, 56.3 years ± 12.1; 42/28 males/females) were retrospectively collected in this study.

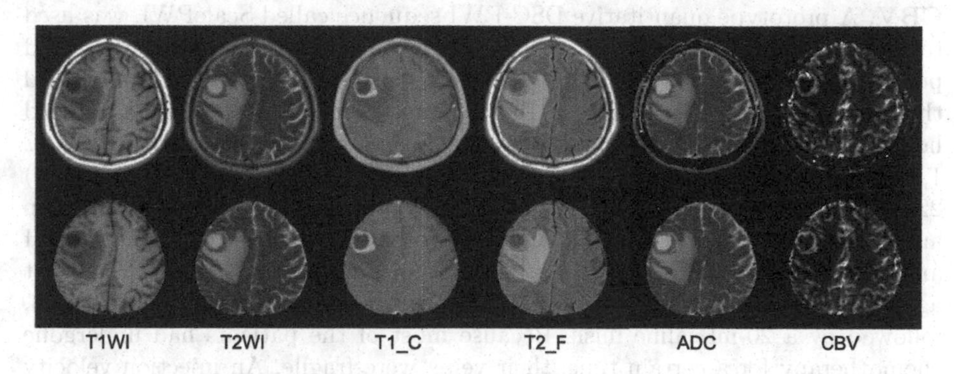

Fig. 1. Illustration of multi-modality MRI sequences (top row) and their segmentation results after skull stripping (bottom row).

2.2 MR Image Acquisition

All patients were imaged in the supine position with a 3.0T MRI machine (Magnetom, Skyra; Siemens Healthcare, Erlangen, Germany) using a transmit/receive quadrature 20-channel head-and-neck coil. The imaging protocol was the same for all patients. In this study, seven imaging modalities were used, including T1WI, T2WI, T2_F, T1_C, ADC map, and CBV map. All axial-view MRI sequences have the same imaging scale, position, slices, and slice thickness, which facilitates co-registration among different modalities.

T1WI TR: 1820 ms, TE: 13 ms, slice number: 19, FOV: 230 mm, slice thickness: 5 mm, distance factor: 30%, FA: 150 deg, inversion time (TI): 825 ms, voxel size: 0.4×0.4×5.0 mm, accelerate factor: 2, bandwidth: 260 Hz/Px, echo spacing: 13 ms.

T2WI Repetition time (TR): 3700 ms, echo time (TE): 109 ms, slices number: 19, field of view (FOV): 220 mm, slice thickness: 5 mm, distance factor: 30%, flip angle (FA): 150 deg, voxel size: 0.3 × 0.3 × 5.0 mm, accelerate factor: 2, bandwidth: 220 Hz/Px, echo spacing: 9.9 ms.

T2_F TR: 8000 ms, TE: 81 ms, slice number: 19, FOV: 220 mm, slice thickness: 5 mm, distance factor: 30%, FA: 150 deg, TI: 2370 ms, voxel size: $0.7 \times 0.7 \times 5.0$ mm, accelerate factor: 2, bandwidth: 289 Hz/Px, echo spacing: 9.02 ms.

ADC TR: 3700 ms, TE1: 65 ms, TE2: 104 ms, slice number: 19, FOV: 230 mm, slice thickness: 5 mm, distance factor: 30%, FA: 180 deg, voxel size: $1.4 \times 1.4 \times 5.0$ mm, accelerate factor: 2, bandwidth: 919 Hz/Px, echo spacing: 0.36 ms, diffusion directions: 3, diffusion mode: 3-Scan trace, diffusion weighting: 2, noise level: 100, b value: 0 and 1000.

T1_C TR: 250 ms, TE: 2.48 ms, slice number: 19, FOV: 230 mm, slice thickness: 5 mm, distance factor: 30%, FA: 70 deg, TI: none, voxel size: $0.7 \times 0.7 \times 5.0$ mm, accelerate factor: 2, bandwidth: 360 Hz/Px, radiofrequency spoiling: yes.

CBV. A prototype quantitative DSC-PWI sequence called ScalePWI, was used to generate CBV map in this study. The ScalePWI sequence merged the pre- and post-contrast T1 mapping into the GRE-EPI sequence for DSC-PWI and added the same "gradient noise" between T1 mapping and the DSC-PWI scan to avoid head motion. The imaging parameters of ScalePWI were as follows: TR/TE 1,600 ms/30 ms, bandwidth 1,748 Hz/pixel, 21 axial slices, field of view (FOV) 220×220 mm, voxel size $1.8 \times 1.8 \times 4.0\,\text{mm}^3$, slice thickness 4.0 mm, and a flip angle (FA) of 90°. For each slice, 50 measurements were acquired for DSC-PWI analysis. After 46 s of injector delay, 0.2 mmol per kg bodyweight of contrast agent (Gd-DTPA, Magnevist; Schering, Berlin, Germany) was administrated, followed by a 20-ml saline flush. Because most of the patients had undergone chemotherapy for a certain time, their veins were fragile. An injection velocity of 4.0 ml/s was introduced in this study.

2.3 Image Processing

All MR images were pre-processed by Statistical Parametric Mapping (SPM) toolbox in MATLAB with four steps: (1) Linearly rigid-align all other volumes (i.e., T1WI, T2WI, T2_F, ADC) to their corresponding T1_C volume. (2) Segment each T1WI brain volume into GM, WM, Cerebro-Spinal Fluid (CSF), skull, scalp and background; (3) The GM, WM, and CSF are conjuncted with a hole-fill morphological operation to create a brain mask; (4) Using the brain mask to remove the non-brain region to let the synthesis algorithms focus on the brain region. After pre-processing, all MR images, a group of which are displayed in Fig. 1, have the same size ($320 \times 320 \times 19$).

2.4 Method

We employ two GAN-based techniques as illustrated in Fig. 2 to explore the possibility to synthesize the CBV map from multiple MR modalities, including the sense-consistency generative adversarial network (SGAN) [12], and the 2D encoder-decoder with only mean absolute error loss (MAE-ED) [2]. We redesigned both these two algorithms to adapt to our dataset, which contains a generative model and a discriminative model.

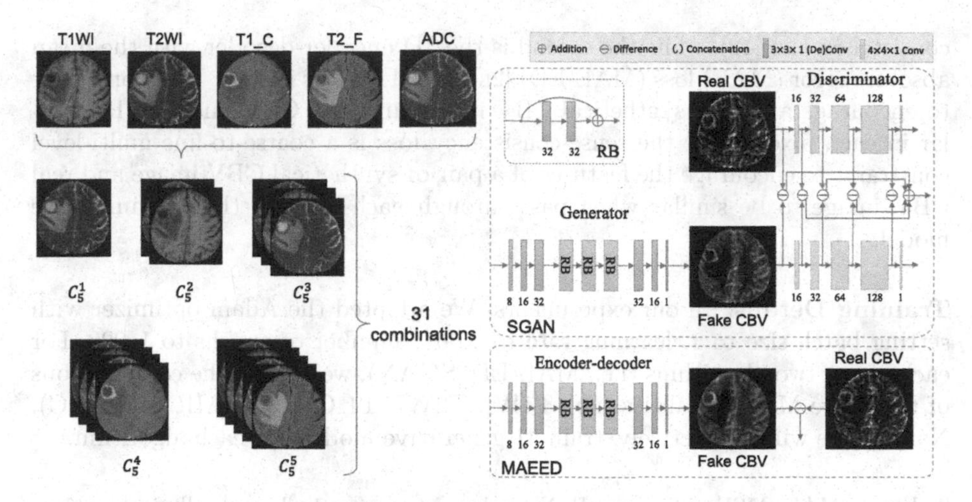

Fig. 2. Description of our investigation in synthesizing CBV maps from multiple MRI modalities. The left is the modality combinations of five MRI sequences while the right are two image synthesis algorithms.

Generative Model. The generative model is composed of sequentially an encoding part, a transferring part and a decoding part. The encoding part uses three convolutional layers (with 16, 32, and 64 channels) to extract the information from the input image(s). The transferring part uses 6 residual network blocks [4] to transfer the information from encoding part to decoding part. The decoding part uses two deconvolutional layers (with 16 and 32 channels) and one convolutional layer (with 1 channel) to construct the target CBV map. All convolutional and deconvolutional layers are with kernel size of $3 \times 3 \times 1$ and followed by instance normalization, which is specifically designed to adapt to our data with large slice thickness. The input of the generative model is the pre-processed multi-modality MRI scans while the output is the CBV map. While inputting to the generative model, the selected MRI scans are concatenated along the channel-axis.

Discriminative Model. The discriminative model consists of 5 convolutional layers, which are 16, 32, 64, 128, and 1 channel(s), respectively. The kernel sizes of these convolutional layers are $4 \times 4 \times 1$ and followed by instance normalization as well. The input of the discriminative model is a synthetical CBV image or a real CBV image while the output is a binary indicator to whether the input image is synthetical or real.

Constraints of Learning. The generative model and discriminative model can form different GANs with different constraints. In this study, we evaluated two GANs based on the above generative model and discriminative model. The first is the sense-consistency GAN (SGAN) with the MAE loss and the sense-consistency loss [12], while the second is the 3D encoder-decoder with the mean absolute error (MAE) loss (MAE-ED) [2]. The MAE loss is a low level constraint to encourage a pair of synthetical CBV image and real CBV image to be similar in each pixel, while the sense-consistency loss is a coarse-to-fine multi-level constraint to encourage the features of a pair of synthetical CBV image and real CBV image to be similar when pass through each layer of the discriminative model.

Training Details. In our experiments, We adopted the Adam optimizer with setting batch size to 1, learning rate to 0.001, number of epochs to 1,000. For each of our two algorithms (i.e., MAE-ED, SGAN), we test all the combinations of these five MRI modalities (i.e., T1WI, T2WI, T1_C, T2_FLAIR, and ADC). Namely, we will, respectively, train 31 generative models for each algorithm.

Table 1. MAE, MSE, SSIM and PSNR values of two methods with all combinations of 5 different modalities.

Combination					SGAN				MAE-ED			
T1WI	T2WI	T1_C	T2_F	ADC	MAE	MSE	SSIM	PSNR	MAE	MSE	SSIM	PSNR
✓					2.648	0.667	80.217	24.302	2.304	0.509	81.411	25.528
	✓				2.448	0.592	79.624	24.912	2.800	0.626	80.498	24.515
		✓			**2.320**	**0.523**	83.501	**25.482**	2.326	0.500	83.228	25.639
			✓		2.815	0.657	80.659	24.342	2.800	0.626	80.498	24.515
				✓	2.611	0.636	81.161	24.538	2.402	0.553	81.215	25.260
✓	✓				2.554	0.612	81.513	24.644	2.348	0.499	82.198	25.557
✓		✓			2.396	0.540	83.093	25.349	2.310	0.493	83.966	25.743
✓			✓		2.621	0.617	81.215	24.643	2.334	0.523	81.487	25.440
✓				✓	2.673	0.657	81.672	24.454	2.589	0.607	82.132	25.031
	✓	✓			2.608	0.608	82.797	24.770	**2.147**	**0.457**	83.608	**26.073**
	✓		✓		2.584	0.613	81.101	24.677	2.712	0.668	80.645	24.856
	✓			✓	2.515	0.606	82.013	24.819	2.244	0.498	82.450	25.647
		✓	✓		2.491	0.570	83.291	25.115	2.263	0.474	**83.986**	25.910
		✓		✓	2.513	0.581	83.468	25.049	2.316	0.525	82.710	25.600
			✓	✓	2.698	0.640	81.005	24.524	2.463	0.545	82.021	25.259
✓	✓	✓			2.490	0.574	82.794	24.992	2.159	0.470	83.081	25.978
✓	✓		✓		2.563	0.593	81.681	24.793	2.386	0.548	81.939	25.383
✓	✓			✓	2.400	0.543	82.028	25.238	2.319	0.506	82.776	25.582
✓		✓	✓		2.637	0.608	82.344	24.749	2.360	0.504	82.905	25.607
✓		✓		✓	2.455	0.556	83.024	25.227	2.178	0.463	83.842	26.039
✓			✓	✓	2.541	0.590	82.073	24.870	2.242	0.541	80.993	25.417
	✓	✓	✓		2.456	0.554	82.951	25.167	2.354	0.548	82.713	25.601
	✓	✓		✓	2.545	0.636	83.331	24.522	2.311	0.532	82.813	25.598
	✓		✓	✓	2.424	0.557	82.313	25.160	2.501	0.581	82.265	25.160
		✓	✓	✓	2.498	0.582	83.382	25.024	2.340	0.492	82.708	25.706
✓	✓	✓	✓		2.605	0.608	82.485	24.784	2.295	0.497	82.397	25.656
✓	✓	✓		✓	2.419	0.548	83.149	25.256	2.178	0.462	83.898	26.050
✓	✓		✓	✓	2.521	0.598	82.040	24.851	2.326	0.515	82.290	25.512
✓		✓	✓	✓	2.566	0.566	82.483	25.093	2.393	0.542	83.593	25.492
	✓	✓	✓	✓	2.488	0.604	**83.791**	24.863	2.293	0.486	83.595	25.793
✓	✓	✓	✓	✓	2.644	0.621	82.801	24.692	2.227	0.478	83.494	25.883

3 Experiments

3.1 Settings and Metrics

The dataset we used contains 70 patients and each patient has one to three groups of data in a following-up visiting time points. Every group contains all these 5 MRI modalities and the calculated CBV map (as an example shown in Fig. 1). 72 groups from 42 patients were used to train our models and the rest 31 groups from 28 patients were used as test data. We use four metrics to evaluate the synthetic CBV maps, including Mean Absolute Error (MAE), Mean Squared Error (MSE), Structural Similarity (SSIM) and Peak Signal-to-Noise Ratio (PSNR).

3.2 Result and Discussion

The experimental results of two different methods on various modality combinations are reported in Table 1. From Table 1, we can find the following observations. *First*, all these combinations seem to be possible to synthesize the CBV maps, which verifies our assumption that other MRI modalities are possible to synthesize CBV maps and it is promising to develop non-invasive and efficient CBV measurement. *Second*, on most of these combinations, MAE-ED achieves slightly better than SGAN in terms of these four metrics, which means different algorithms will affect the performance, thus more effective algorithms should be explored in the further. *Third*, using more modalities does not general achieve better performance then using less modalities for either algorithm in any metric, which suggests that some modalities may be not necessary to acquire. *Fourth*, using all modalities does not achieve the best performance but also not the worst performance, which may be a balanced trade-off. But considering the acquisition time, it may be not the best choice. *Fifth*, the best combination is different for different algorithms and different metrics. For example, the best MAE, MSE, and PSNR values (which are 2.147, 0.457, and 26.073 respectively) for the synthetic results of MAE-ED are achieved by the combination of T1_C and T2WI, while the best MAE, MSE and PSNR values of SGAN, (which are 2.320, 0.523, and 25.482, respectively) are achieved by the combination of T1_C only. It suggests that different algorithms may have biased focus on different aspects. *Sixth*, the best performance of SGAN is achieved by using only T1_C over all metrics, while of MAE-ED is achieved by using T2WI and T1_C over MAE, MSE, and PSNR and using T1_C and T2_F over SSIM. It suggests that T1_C is the most helpful modality in synthesizing CBV maps.

Figure 3 displays an example of our two algorithms (i.e., MAE-ED, SGAN) on all modality combinations. From Fig. 3, two additional observations can be found. *First*, on some combinations, the synthetic CBV maps are relatively darker than the real map. This is due to the fact that the original images have a large intensity diversity which has not been well suppressed in our pre-processing stage. This should be avoided in our further work. *Second*, these maps synthesized by MAE-ED are smoother than those synthesized by SGAN. It is because

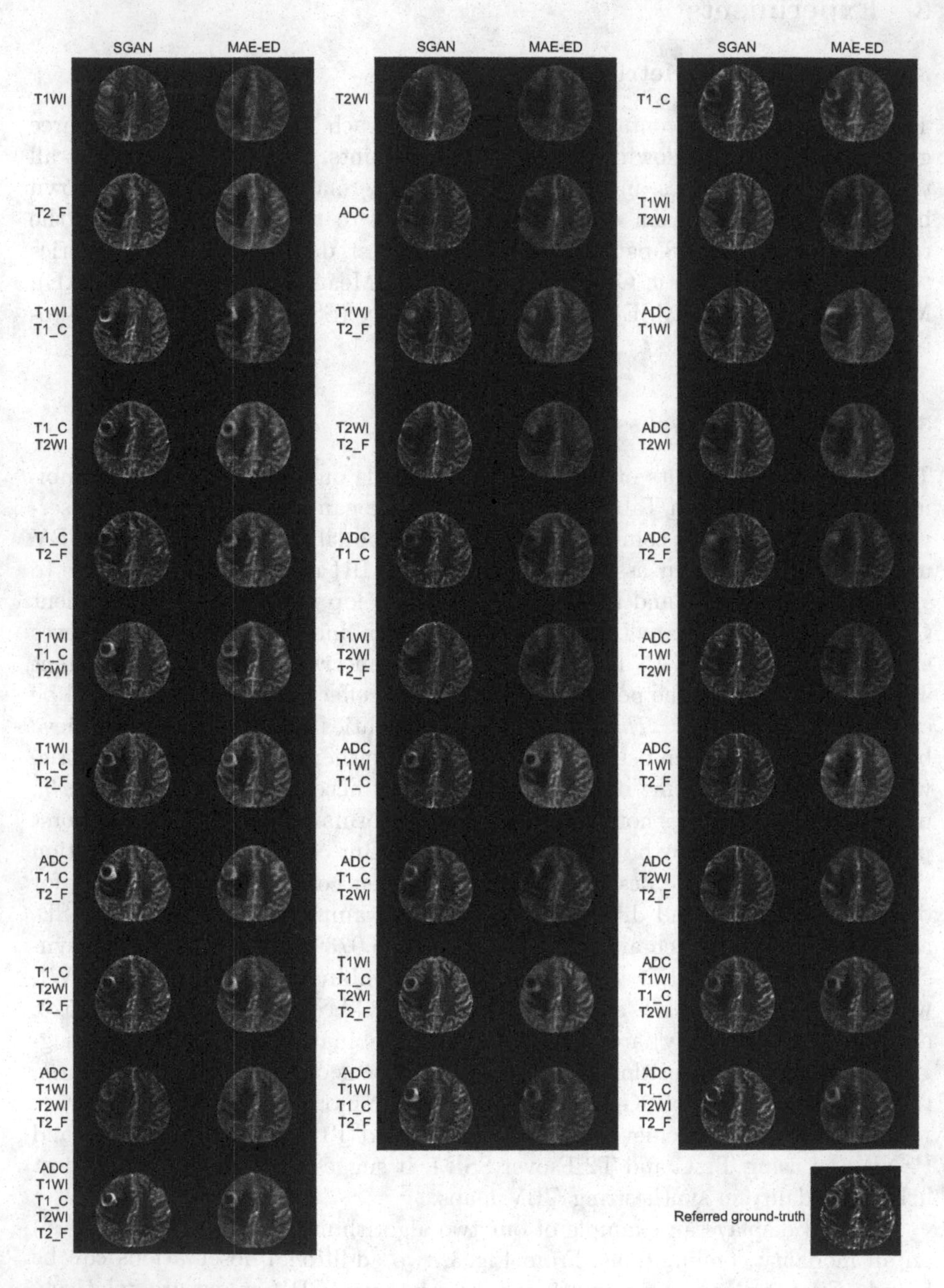

Fig. 3. An example of our two algorithms to synthesize CBV map while using different modality combinations. The left and right are the synthetic results of SGAN and MAE-ED, respectively. The corresponding ground-truth is in the bottom-right.

the MAE loss have a smooth effect, which is also a phenomenon in some previous work [6,11,12]. By contrast, SGAN has a better visual effect in synthesizing CBV maps.

4 Conclusion

In this paper, we collected 103 groups of multiple MRI modalities of 70 subjects to investigate the possibility of synthesizing the CBV maps from multiple-modality MR images, i.e., T1WI, T2WI, T1_C, T2_F, and ADC. We reimplemented the sense-consistency generative adversarial network (SGAN) and the 3D encoder-decoder with only mean absolute error loss (MAE-ED) to apply them to our dataset. The experimental results on various combinations of these five modalities demonstrate that our SGAN and MAE-ED is possible to synthesize promising CBV maps, which may be useful to reduce the patients' discomfort and the diagnosis cost. Our future work will focus on improving our algorithm to synthesize more reliable CBV maps.

Acknowledgment. This work was supported in part by the Science and Technology Innovation Committee of Shenzhen Municipality, China, under Grants JCYJ20180306171334997, in part by the National Natural Science Foundation of China under Grants 61771397, in part by the China Postdoctoral Science Foundation under Grants BX2021333, and in part by the Taishan Scholars Program under Grants tsqn20161070.

References

1. Cercignani, M., Bouyagoub, S.: Brain microstructure by multi-modal MRI: is the whole greater than the sum of its parts? Neuroimage **182**, 117–127 (2018)
2. Cohen, J.P., Luck, M., Honari, S.: Distribution matching losses can hallucinate features in medical image translation. In: Frangi, A.F., Schnabel, J.A., Davatzikos, C., Alberola-López, C., Fichtinger, G. (eds.) MICCAI 2018. LNCS, vol. 11070, pp. 529–536. Springer, Cham (2018). https://doi.org/10.1007/978-3-030-00928-1_60
3. Donahue, M.J., Juttukonda, M.R., Watchmaker, J.M.: Noise concerns and post-processing procedures in cerebral blood flow (CBF) and cerebral blood volume (CBV) functional magnetic resonance imaging. Neuroimage **154**, 43–58 (2017)
4. He, K., Zhang, X., Ren, S., Sun, J.: Deep residual learning for image recognition. In: Proceedings of the IEEE Conference on Computer Vision and Pattern Recognition, pp. 770–778 (2016)
5. Hua, J., et al.: MRI techniques to measure arterial and venous cerebral blood volume. Neuroimage **187**, 17–31 (2019)
6. Isola, P., Zhu, J.Y., Zhou, T., Efros, A.A.: Image-to-image translation with conditional adversarial networks. In: Proceedings of the IEEE Conference on Computer Vision and Pattern Recognition, pp. 5967–5976 (2017)
7. Jilaveanu, L.B., et al.: Tumor microvessel density as a prognostic marker in high-risk renal cell carcinoma patients treated on ECOG-ACRIN E2805. Clin. Cancer Res. **24**(1), 217–223 (2018)
8. Laviña, B.: Brain vascular imaging techniques. Int. J. Mol. Sci. **18**(1), 70 (2017)

9. Lee, F.K.H., King, A.D., Ma, B.B.Y., Yeung, D.K.W.: Dynamic contrast enhancement magnetic resonance imaging (DCE-MRI) for differential diagnosis in head and neck cancers. Eur. J. Radiol. **81**(4), 784–788 (2012)
10. Ma, H., et al.: Three-dimensional arterial spin labeling imaging and dynamic susceptibility contrast perfusion-weighted imaging value in diagnosing glioma grade prior to surgery. Exp. Ther. Med. **13**(6), 2691–2698 (2017)
11. Pan, Y., Liu, M., Lian, C., Xia, Y., Shen, D.: Spatially-constrained fisher representation for brain disease identification with incomplete multi-modal neuroimages. IEEE Trans. Med. Imaging **39**(9), 2965–2975 (2020)
12. Pan, Y., Xia, Y.: Ultimate reconstruction: understand your bones from orthogonal views. In: Proceedings of the IEEE International Symposium on Biomedical Imaging, pp. 1155–1158. IEEE (2021)
13. Shao, B., Liu, E.: Expression of ING4 is negatively correlated with cellular proliferation and microvessel density in human glioma. Oncol. Lett. **14**(3), 3663–3668 (2017)

Cine-MRI Simulation to Evaluate Tumor Tracking

José D. Tascón-Vidarte[1(✉)], Isak Wahlstedt[2,3], Julien Jomier[4], Kenny Erleben[1], Ivan R. Vogelius[2], and Sune Darkner[1]

[1] Department of Computer Science, University of Copenhagen, Copenhagen, Denmark
{jota,kenny,darkner}@di.ku.dk
[2] Department of Oncology, Rigshospitalet, Copenhagen, Denmark
{isak.hannes.wahlstedt,ivan.richter.vogelius}@regionh.dk
[3] Technical University of Denmark, Kongens Lyngby, Denmark
[4] Kitware SAS, Villeurbanne, France
julien.jomier@kitware.com

Abstract. Conventional evaluations of tumor tracking algorithms require inter-observer segmentations from radiation oncologists on the Cine-MRI (2D sagittal MR video). Instead of performing intensive manual annotations on images, we present a 2D video simulator that uses the pre-treatment images, including a breathing model, that generates Cine-MR images in parallel with the underlined segmentation of the tumor. We include the data of seven patients within a retrospective clinical study that received stereotactic body radiation therapy for liver metastases. Each patient has a pre-treatment 4DCT scan, a pre-treatment 3D MR with tumor and liver delineations, and the treatment Cine-MRI. We augment the data with the simulator by changing breathing motion parameters and adding noise. The simulator generates a total of 84 Cine-MRI sequences, thus having 12 videos per patient. We validate the simulated versus the real Cine-MRI in terms of tumor motion. Finally, we used the simulator to evaluate the performance of real-time tumor tracking algorithms with this dataset.

Keywords: Cine-MRI · Simulation · Tumor tracking · Real-time · Image-guided radiotherapy

1 Introduction

An MR-linac is a device that combines magnetic resonance imaging with a linear accelerator. Stereotactic body radiotherapy (SBRT) of liver metastases on the MR-Linac system is advantageous due to the improved soft-tissue contrast compared to cone-beam computed tomography [12]. In addition, the MR-Linac has beam gating, i.e. the treatment accelerator beam is triggered in response to patient movement [4]. The clinical advantages of using an MR-linac with online

D. Svoboda et al. (Eds.): SASHIMI 2021, LNCS 12965, pp. 131–141, 2021.
https://doi.org/10.1007/978-3-030-87592-3_13

tracking have been demonstrated [1], indicating the potential to reduce the liver target volume and lower the radiation dose to adjacent organs at risk.

Liver tumors deform and move during treatment mainly caused by breathing motion [14]. During treatment, the scanner acquires 2D cine-MR sagittal images at four (4) frames per second [16]. Tumor tracking is one of the main components of beam gating in the MR-Linac system. Tumor tracking is solved automatically using image analysis. Some proposed strategies for tumor tracking are based on template matching [3,19], feature detection [15], optical-flow methods [17, 23], deformable image registration [5], segmentation [9], neural networks [22] or modeling based [8]. The difficulty in evaluating tumor tracking is the need for ground truth data as no publicly available datasets that comprise tumor tracking on Cine-MRI exist. This requires manually delineation of the tumor in the entire video by a radiation oncologist [3,5,19,22]. Evaluations are therefore only comparable at the institutional level. A generalized methodology to easily evaluate tumor tracking is required.

In addition, current tracking systems used in clinical practice may fail to track unexpected movements and have difficulty in tracking motion in the out-of-plane direction [16]. The breath-hold treatment is the most used respiratory motion management in practice [6]. After breath-hold, the patients can exhibit a fast motion, and thus tracking becomes very difficult. All the previously reported studies evaluate tumor tracking under free-breathing conditions [3,5,8,15,17, 19,22,23]. Thus, there is a need for improved tracking algorithms under varying breathing motion.

Our contribution is a straightforward evaluation methodology to quantify tumor tracking performance without the need of manual segmentations. The method is patient-specific and simple to implement. We demonstrate the capabilities of our method to create multiple simulated Cine-MRI, and to evaluate tumor tracking algorithms under varying conditions.

2 Related Work

Respiratory motion modeling is an extensively studied field [13]. Deformable image registration generates the most suitable models [20]. Likewise, our breathing model works with deformable image registration. Fu et al. [7] use known deformation fields to create ground truth images and landmarks to validate feature detection on 4DCT. Our simulator works in a similar way, but in contrast the known transformation model is applied to the images and the tumor contours.

The use of pre-treatment 4D imaging data has been exploited for treatment. Harris et al. [11] use 4DMRI to create synthetic 4DCT during treatment on conventional linear accelerators. Garau et al. [8] use pre-treatment 4DCT with treatment Cine-MRI to estimate a 3DCT and compare planning versus treatment. The mentioned methods suggest multi-modal integration of images and that pre-treatment time-sequence images are valid to model breathing motion during treatment. We aim for a similar goal with a distinction, to build a breathing model based on registration, use the model to simulate treatment sequences, in order to improve tumor tracking algorithms. To our knowledge this is the first Cine-MRI simulation based on a respiratory motion model.

An alternative option to manually delineate the tumor is to use matched landmarks [15,17]. Although this process can be automated with feature detection algorithms some outliers occur. Furthermore, the landmark distance alone does not represent how well the tracking algorithm performs with respect to the tumor structure and the contours. Most tumor tracking algorithms have been tested on lung patients with manual delineations of the tumor [3,5,19,22] and only a small set has been evaluated on liver patients with landmarks [15,17]. We test the proposed methodology and the tracking algorithms with liver patients. Identifying a tumor contour on a sagittal slice is difficult in the liver due to the lack of contrast. Hence, the liver is a remarkable example of why manual tumor delineations for evaluating tumor tracking are not always feasible.

3 Methods

3.1 Data

This study uses image data from seven patients already treated with SBRT for metastases in the liver at Rigshospitalet (Copenhagen, Denmark) between April and December 2019. The patients provided informed consent and approval for the usage of their anonymized data for research purposes.

Respiratory correlated 4DCT with intravenous contrast injection was performed for all patients on a SOMATOM Definition AS scanner (Siemens Healthineers, Germany). 4DCT image data were phase-sorted into ten phase bins throughout a respiratory cycle based on an external respiratory signal monitored with Real-Time Position Management (RPM, Varian Medical Systems, USA). The slice separation in each phase of the 4DCT was 2 mm. The image resolution in each slice was 512×512 pixels and a pixel size of 0.98×0.98 mm.

A 0.3T MRIDIAN MR-Linac (ViewRay, USA) is used to acquired a pre-treatment 3D MRI and the Cine-MRI sequences. The 3D MR scans were performed for all patients in inspiration breath-hold position without visual guidance. The acquisition technique is balanced steady-state free precession (bSSFP). The pre-treatment image resolution is [$512 \times 512 \times 128$ pixels] and [$1.5 \times 1.5 \times 3.0$ mm] spacing. The Cine-MRI sequences (bSSFP-Sagittal) have a resolution of [256×256 pixels] and [1.5×1.5 mm] spacing. The clinical gross tumor volume was delineated on the 3D MR by a senior radiologist and approved by a senior oncologist. Organs at risk, including the liver, were also delineated. These delineated contours are used to segment the region of interest (ROI).

3.2 Cine-MRI Simulation

We developed a patient-specific Cine-MRI simulator capable to generate a simulated ground truth contour of the desired organs using pre-treatment images. The input images are a 4DCT scan and an MR with organ contours. The video simulator has the following input parameters: video time, frames per second, breathing cycle time, breathing amplitude, and additive noise. Figure 1 illustrates the video simulation process. Algorithm 1 details the simulator pseudo code. The simulator has two stages: breathing modeling and video synthesis. Each stage is described in the following.

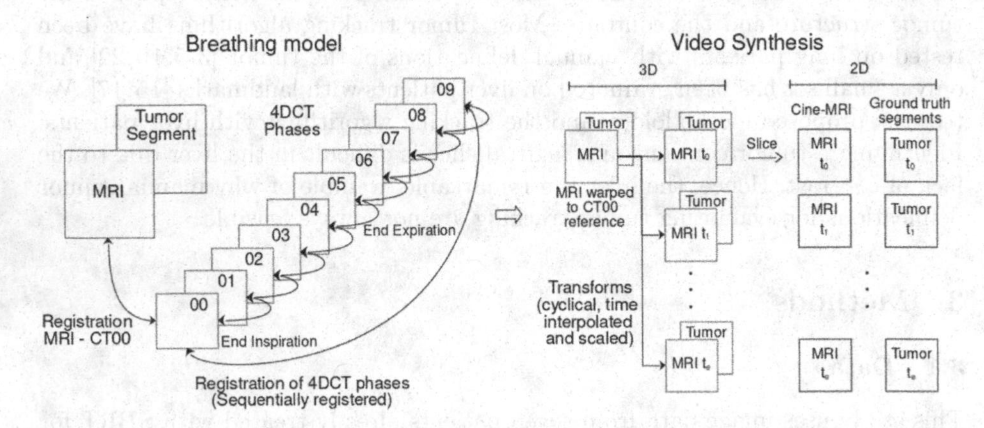

Fig. 1. 2D Cine-MRI Simulator. The process comprises two stages: breathing modeling and video synthesis. The breathing model is a pre-processing stage and uses full 3D information to consider out-of-plane motion in the 2D Cine-MRI. The video synthesis stage can be run several times changing the simulation parameters to create different variants and motion conditions.

The breathing modeling is a pre-processing stage. It is computed once and stored in order to create several videos. This model is based on the 4DCT scan that represents the full respiratory cycle of each patient. Initially, all phases in the 4DCT are registered sequentially with the symmetric normalization algorithm [2]. Subsequently, the MR image is registered to phase 00 of the 4DCT scan, since both images are at inspiration position. This transformed MR is the starting video frame.

The video synthesis stage is an iterative stage. The simulator produces a new video frame as a composition of sequential transformations related to the 4DCT. The corresponding transformation is interpolated over time to match the proportion of the respiratory cycle with the video sample time. The video is created based on 3D images and transformations. From this a 2D slice is extracted in the sagittal view where the tumor has better visibility. Thus the simulated video has the same complexity as real 3D motion in the 2D images and simulates the MR-Linac imaging setup, where 2D real-time images are acquired and tracked. Since the initial contour of the tumor and organs are known in the reference MR, we create independent files with those regions of interest (ROI) and the ROI are transformed in parallel with the raw video image generation. For simplicity we only generate the tumor contours. Therefore, we have the ground truth ROI for each video sample.

Algorithm 1. Cine-MRI Simulation

Input: 4DCT, 3D MRI (bSSFP) and clinical tumor contour
Parameters: Video time, frames per second, breathing cycle time, breathing amplitude, and additive noise
Output: Cine-MRI (Video containing 2D bSSFP-Sagittal images)

1: Register sequentially 4DCT images, and store transformations $\varphi_{00\to01}, \varphi_{01\to02}, ..., \varphi_{09\to00}$ where $j = 00, 01, ..., 09$ correspond to the phase;
2: Register MRI image to 4DCT phase 00 obtain $\varphi_{MR\to00}$;
3: Transform MRI image to CT phase 00 using $\varphi_{MR\to00}$, obtain M_{00};
4: Transform tumor delineation X_{MR} using $\varphi_{MR\to00}$ to get contour X_{00};
5: Compute simulation *time* using parameters: video time, frames per second;
6: **for** t in *time* **do**
7: Compute time point t location in breathing cycle;
8: Calculate time proportion for time interpolation t_δ;
9: Find the shortest path of sequential transformations;
10: Compose a transformation $\varphi_c(x)$ with the shortest path;
11: Multiply $\varphi_c(x)$ with breathing amplitude parameter α;
12: Transform image M_{00} using $\varphi_c(x)$, obtain M_t;
13: Extract a slice of M_t to create the image $M_{2d,t}$;
14: Add noise to image $M_{2d,t}$;
15: Transform contour X_{00} using $\varphi_c(x)$, obtain X_t;
16: Extract a slice of X_t to create the image $X_{2d,t}$;
17: **end for**

The simulator supports noise with two different probability distributions: Gaussian and Rician distributions. Noise in MR images is often modeled as Rician, and for signal to noise ratio greater than two, the noise behaves like Gaussian [10]. Furthermore, other sources of noise from the MR device are still modelled as Gaussian.

3.3 Tracking Algorithms

The MR-Linac typical rate of acquisition is four (4) frames per second [16]. Tracking algorithms must meet this time requirement. The first image and its corresponding tumor contour in the Cine-MR sequence is used as the reference and the tracking algorithms uses the subsequent Cine-MRI input images to estimate a new tumor contour. Fast et al. [5] presented a comparative study where they analyzed four tumor tracking techniques. The authors concluded that all the algorithms had a relatively similar performance but among them deformable image registration and template matching provided slightly better results. We implemented these two algorithms to evaluate tumor tracking simulator. Both algorithms are implemented on C++ and parallelized on CPU with OpenMP.

We chose the diffeomorphic demons [21] as a fast solution of deformable image registration. Our approach uses a multiresolution framework with three pyramidal levels. The computational bottleneck of registration is the computation of the transformation and the similarity metric [18]. We focused on this stage to improve performance. Regarding the optimization, we fix the iteration values to be able to achieve the time restriction, this means that full convergence is not always guaranteed.

We implemented the generalized template matching algorithms [3,19]. The algorithm defines a template in a reference image, usually the tumor bounding box, and search for it in a local region of the input image by maximizing an objective function to determine a good match. The preferred function is cross-correlation for Cine-MR tracking. The wider the search region is, the more computational time, while with a more limited search region, there is a risk of not capturing the tumor motion.

3.4 Metrics

We used three metrics to validate the real and simulated Cine-MRI and to evaluate the tumor tracking algorithms. First, the Dice Similarity Coefficient, which serves to quantify the whole segmented structure. Secondly, the centroid distance, which quantify the algorithm's ability to follow the center of mass (COM) of the tumor. And third, the Hausdorff distance, which provides a measurement of the effectiveness to detect and track the tumor contours. Further information regarding the metrics are detailed in Fast et al. [5].

4 Experiments and Results

4.1 Patient Summary

Table 1 summarizes the patients tumor and breathing motion. The patient set is small in numbers but represents a wide variety of anatomical tumor location, tumor size, breathing motion, and breathing cycle times. The most challenging conditions for the tracking algorithms are a short tumor displacement or a small tumor area.

Table 1. Summary of patient information. Tumor location is the geometric octant of where the tumor is with regards to the liver center of mass. The abbreviations correspond to Superior-Inferior, Anterior-Posterior, and Left-Right. Breathing cycle times were determined from real patient respiratory motion during 4DCT scans. Tumor displacements refer to the maximum motion presented in the video without registration. Gross tumor volumes/areas are estimated on the reference 3D/2D (sagittal) MR.

Patient	1	2	3	4	5	6	7
Tumor location (geometric)	S-A-L	S-A-L	S-A-R	I-A-L	S-A-R	S-P-L	I-P-R
Breathing cycle (mean) [s]	4.8	3	4.1	3.6	4.4	4.1	5.8
Tumor max. displacement [mm]	5.9	9.1	8.1	9.7	2.0	6.5	12.6
Tumor volume [cm^3]	5.5	5.6	12.0	3.1	3.1	8.0	5.7
Tumor sagittal area [cm^2]	2.7	3.0	7.4	3.3	1.8	4.7	4.1

4.2 Cine-MRI Image Quality

Figure 2 shows a comparison between the real Cine-MRI and the simulation for one patient. All the resulting videos are visually close to the real Cine-MRI. To validate the real versus the simulated Cine-MRI, we verify how consistent the tumor motion is for all the patients. We select 12 images (approximately a breathing cycle) of the real Cine-MRI with the first image in the inspiration position. The tumors are segmented manually on the real Cine-MRI. Since the real and the simulated data have a slightly different field of view, we perform rigid registration around the tumor on the first image and align the remaining simulation images using the same transformation. We calculate the metrics for all the patients with the real versus the simulated Cine-MRI. The Dice score is 0.89 ± 0.05 (*mean* ± *std.dev.*), the centroid distance 0.78 ± 0.32 mm and Hausdorff distance 2.11 ± 0.91 mm.

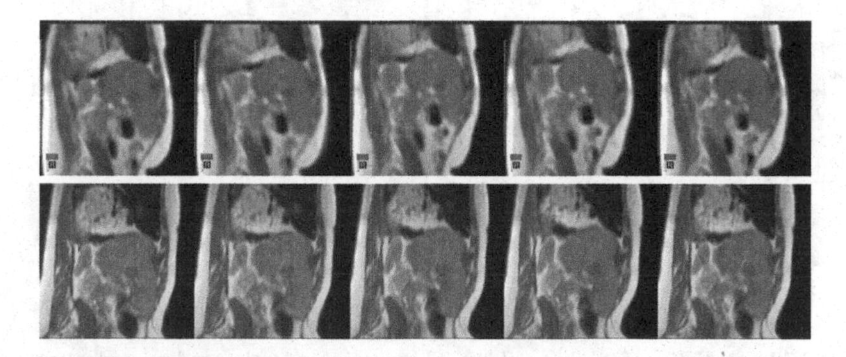

Fig. 2. Video comparison of real Cine-MRI and simulated of one patient. Top images depict the real Cine-MRI and the bottom images the simulation.

4.3 Tumor Tracking Performance

We evaluate tumor tracking with a full factorial experiment between breathing amplitude and noise. The breathing amplitude is varied with values 1.0, 1.5, 2.0, or random. The noise is varied between none, Gaussian and Rician. Gaussian noise and Rician noise applied in all the tests are equivalent to 20% of added noise. These experiments generate 12 videos per patient for a total of 84 Cine-MRI sequences. The breathing cycle parameter is patient-specific taken from Table 1. All the videos are 20 s long at 4 frames per second (80 images), approximately 4–5 breathing cycles.

Figure 3 depict the tumor tracking performance. For comparison, we compute a baseline (in blue) that corresponds to the metric value without tracking. Videos 1 to 4 vary in amplitude without noise, videos 5 to 8 vary in amplitude with Gaussian noise, and videos 9 to 12 vary in amplitude with Rician noise. Video 1 is the most representative as it has the default and control conditions. The

Dice score summary as $mean \pm standarddeviation$ results of registration are 0.88 ± 0.06, template matching 0.79 ± 0.12. The centroid distance obtained for the registration is 0.89 ± 0.54 mm versus template matching 1.71 ± 3.81 mm. The Hausdorff distance obtained for registration is 3.23 ± 1.35 mm versus template matching 4.41 ± 4.02 mm. Both algorithms' performance in terms of centroid distances is adequate for image-guided radiotherapy.

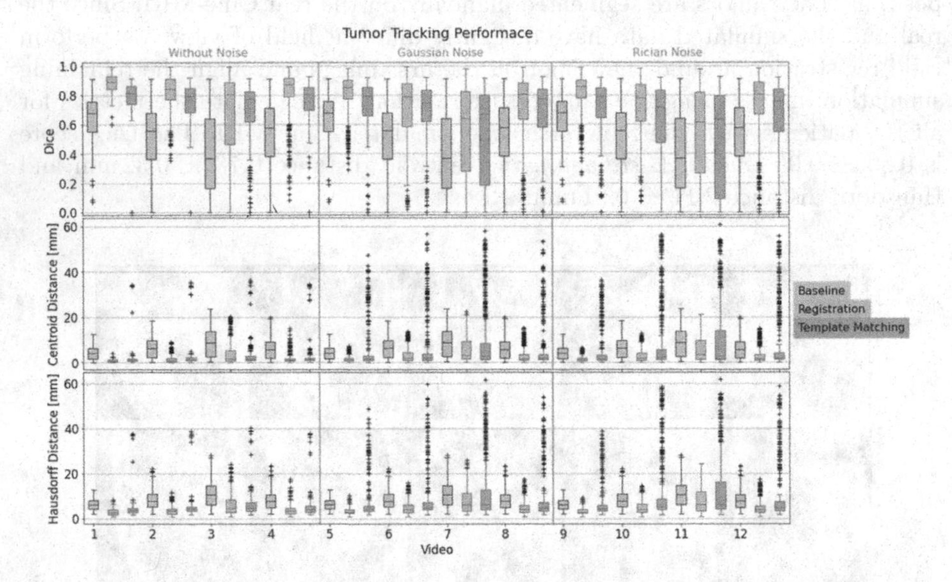

Fig. 3. Video statistics of tumor tracking. The metrics are estimated per video and comprise all patients. Videos are generated as a full factorial experiment between breathing amplitude and added noise. Ascending numbers in groups of 4 videos correspond to breathing amplitude of 1, 1.5, 2, and random respectively. The metrics are shown from top to bottom as Dice, centroid distance, and Hausdorff distance. In blue, the baseline as the metrics computed without tracking. In green, the metrics determined with registration. In red, the metrics estimated with template matching. (Color figure online)

All the tests were run on a workstation with 2 CPUs and 128 GB of RAM. Each CPU is an Intel(R) Xeon(R) Silver 4110 @ 2.10 GHz, 8 cores, 16 threads. The computational time of the deformable registration algorithm time is on average 62.7 ms with a standard deviation of 42.3 ms. The maximum registration time is 242 ms. The computational time of the template matching algorithm is on average 16.7 ms with a standard deviation of 4.4 ms.

5 Discussion and Conclusion

We validate the tumor motion between the real versus the simulated Cine-MRI. The centroid distance is the best metric to represent the motion, and its mean value distance of 0.78 mm indicates a high similarity between the real and the simulated data. We do not compare image intensities between the real and the simulated Cine-MRI because they come from sources acquired on different dates and under different conditions (pre-treatment - treatment, the field of view, alignment, among others). Furthermore, a direct comparison of pixel intensities or image similarity will only reveal how close they are in terms of signal, contrast, or even alignment but not how well the simulation model breathing and tumor motion.

Regarding our specific evaluation of tumor tracking algorithms, we identify in general that deformable image registration perform better. The template matching algorithm fails under noisy conditions and present several outliers. A breathing amplitude of 2.0 is an extreme condition and unrealistic. However, from the algorithms point of view is an interesting experiment. Both tracking algorithms fails to follow the tumor having wide ranges of Dice scores under this condition.

A limitation of our Cine-MRI simulator is that the breathing model uses a single respiratory cycle from the 4DCT scan. The breathing model overcomes this by composing transformations that are time interpolated. Time interpolation guarantees that different patterns arise due to asynchrony between the patient's breathing cycle time and sampling times. Furthermore, when we model with full 3D images and then create the 2D Saggital MR, we incorporate the desired out-of-plane motion, which is the main challenge for tracking algorithms. Overall, our goal is not to create a perfect breathing model but to facilitate challenging experiments to evaluate tumor tracking algorithms.

We designed a platform and a methodology to easily evaluate tracking algorithms on Cine-MR with ground truth segmentation. The video simulator does not require any training data and works only with pre-treatment images. The proposed methodology is the most automated way to evaluate tumor tracking algorithms with a ground truth. Our code is open source and available at https://github.com/josetascon/cinemri-simulation.

Acknowledgment. This project has received funding from the European Union's Horizon 2020 research and innovation program under the Marie Sklodowska-Curie grant agreement No. 764644. This paper only contains the author's views and the Research Executive Agency and the Commission are not responsible for any use that may be made of the information it contains.

References

1. Al-Ward, S., et al.: The radiobiological impact of motion tracking of liver, pancreas and kidney SBRT tumors in a MR-linac. Phys. Med. Biol. **63**(21), 215022 (2018)
2. Avants, B.B., Epstein, C.L., Grossman, M., Gee, J.C.: Symmetric diffeomorphic image registration with cross-correlation: evaluating automated labeling of elderly and neurodegenerative brain. Med. Image Anal. **12**(1), 26–41 (2008)
3. Cervino, L.I., Du, J., Jiang, S.B.: MRI-guided tumor tracking in lung cancer radiotherapy. Phys. Med. Biol. **56**(13), 3773 (2011)

4. Crijns, S., Kok, J., Lagendijk, J., Raaymakers, B.: Towards MRI-guided linear accelerator control: gating on an MRI accelerator. Phys. Med. Biol. **56**(15), 4815 (2011)
5. Fast, M.F., et al.: Tumour auto-contouring on 2d cine MRI for locally advanced lung cancer: a comparative study. Radiother. Oncol. **125**(3), 485–491 (2017)
6. Feldman, A.M., Modh, A., Glide-Hurst, C., Chetty, I.J., Movsas, B.: Real-time magnetic resonance-guided liver stereotactic body radiation therapy: an institutional report using a magnetic resonance-linac system. Cureus, **11**(9) (2019)
7. Fu, Y., Wu, X., Thomas, A.M., Li, H.H., Yang, D.: Automatic large quantity landmark pairs detection in 4dct lung images. Med. Phys. **46**(10), 4490–4501 (2019)
8. Garau, N., et al.: A ROI-based global motion model established on 4DCT and 2D cine-MRI data for MRI-guidance in radiation therapy. Phys. Med. Biol. **64**(4), 045002 (2019)
9. Gou, S., Wu, J., Liu, F., Lee, P., Rapacchi, S., Hu, P., Sheng, K.: Feasibility of automated pancreas segmentation based on dynamic MRI. Br. J. Radiol. **87**(1044), 20140248 (2014)
10. Gudbjartsson, H., Patz, S.: The Rician distribution of noisy MRI data. Magn. Reson. Med. **34**(6), 910–914 (1995)
11. Harris, W., Wang, C., Yin, F.F., Cai, J., Ren, L.: A novel method to generate on-board 4D MRI using prior 4D MRI and on-board KV projections from a conventional linac for target localization in liver SBRT. Med. Phys. **45**(7), 3238–3245 (2018)
12. Kontaxis, C., Bol, G., Stemkens, B., Glitzner, M., Prins, F., Kerkmeijer, L., Lagendijk, J., Raaymakers, B.: Towards fast online intrafraction replanning for free-breathing stereotactic body radiation therapy with the MR-linac. Phys. Med. Biol. **62**(18), 7233 (2017)
13. McClelland, J.R., Hawkes, D.J., Schaeffter, T., King, A.P.: Respiratory motion models: a review. Med. Image Anal. **17**(1), 19–42 (2013)
14. Murphy, M.J.: Tracking moving organs in real time. In: Seminars in Radiation Oncology, vol. 14, pp. 91–100. Elsevier (2004)
15. Paganelli, C., et al.: Magnetic resonance imaging-guided versus surrogate-based motion tracking in liver radiation therapy: a prospective comparative study. Int. J. Radiat. Oncol*. Biol*. Phys. **91**(4), 840–848 (2015)
16. Paganelli, C., et al.: MRI-guidance for motion management in external beam radiotherapy: current status and future challenges. Phys. Med. Biol. **63**(22), 22TR03 (2018)
17. Seregni, M., Paganelli, C., Summers, P., Bellomi, M., Baroni, G., Riboldi, M.: A hybrid image registration and matching framework for real-time motion tracking in MRI-guided radiotherapy. IEEE Trans. Biomed. Eng. **65**(1), 131–139 (2017)
18. Shams, R., Sadeghi, P., Kennedy, R.A., Hartley, R.I.: A survey of medical image registration on multicore and the GPU. IEEE Signal Process. Mag. **27**(2), 50–60 (2010)
19. Shi, X., Diwanji, T., Mooney, K.E., Lin, J., Feigenberg, S., D'Souza, W.D., Mistry, N.N.: Evaluation of template matching for tumor motion management with cine-MR images in lung cancer patients. Med. Phys. **41**(5), 052304 (2014)
20. Stemkens, B., Tijssen, R.H., De Senneville, B.D., Lagendijk, J.J., Van Den Berg, C.A.: Image-driven, model-based 3d abdominal motion estimation for MR-guided radiotherapy. Phys. Med. Biol. **61**(14), 5335 (2016)
21. Vercauteren, T., Pennec, X., Perchant, A., Ayache, N.: Diffeomorphic demons: efficient non-parametric image registration. Neuroimage **45**(1), S61–S72 (2009)

22. Yun, J., Yip, E., Gabos, Z., Wachowicz, K., Rathee, S., Fallone, B.: Neural-network based autocontouring algorithm for intrafractional lung-tumor tracking using linac-MR. Med. Phys. **42**(5), 2296–2310 (2015)
23. Zachiu, C., Papadakis, N., Ries, M., Moonen, C., De Senneville, B.D.: An improved optical flow tracking technique for real-time MR-guided beam therapies in moving organs. Phys. Med. Biol. **60**(23), 9003 (2015)

GAN-Based Synthetic FDG PET Images from T1 Brain MRI Can Serve to Improve Performance of Deep Unsupervised Anomaly Detection Models

Daria Zotova[1(✉)], Julien Jung[2], and Carole Lartizien[1]

[1] Univ Lyon, CNRS, Inserm, INSA Lyon, UCBL, CREATIS, UMR5220, U1206, 69621 Villeurbanne, France
{Daria.Zotova,Carole.Lartizien}@creatis.insa-lyon.fr

[2] Lyon Neuroscience Research Center, CRNL, INSERM U1028, CNRS UMR5292, University Lyon 1, Lyon, France

Abstract. Research in cross-modal translation or synthesis domain has been very productive over the past few years to tackle the scarce availability of large curated datasets for the training of deep models, with promising performance of GAN-based architectures. However, only a few of these studies assessed task-based related performance of these synthetic data. In this work, we design and compare different GAN-based frameworks for generating synthetic brain FDG-PET images from T1-weighted MRI data, and explore further impact of adding these fake PET data in the training of a deep brain anomaly detection model. Qualitative and quantitative results allow us to conclude that the generated PET images look similar to real ones with SSIM and PSNR values around 0.88 and 23.5 respectively for the best GAN architecture. Training of the brain anomaly detection model on hybrid datasets including 35 real and 40 synthetic FDG PET data, allows achieving a 65% detection sensitivity of subtle epilepsy lesions in 17 real PET exams of patients, while the sensitivity is 53% when training with the 35 real PET exams only, thus demonstrating the diagnostic value of these synthetic data for the design of CAD models.

Keywords: Medical image synthesis · Unsupervised learning · Cycle-GAN · PET MRI · Lesion detection

1 Introduction

One major limitation to the performance of deep learning models for medical image analysis is the scarce availability of large annotated training databases. It is all the more difficult to acquire large datasets of paired multi-modality exams (we indeed often have to deal with missing or incomplete datasets) and to sample the variability of the normal and pathological pattern distributions. For this

D. Svoboda et al. (Eds.): SASHIMI 2021, LNCS 12965, pp. 142–152, 2021.
https://doi.org/10.1007/978-3-030-87592-3_14

reason, unsupervised or weakly supervised paradigms have gained significant interest over the past few years, due the constraint release on the annotation process [11]. This includes anomaly detection models, which were shown to perform well, especially for detection and segmentation tasks in neuroimaging [3]. This subgroup of methods which consists in learning normal representations or patterns extracted from normal (i.e. non pathological) populations only, allow relaxing the pressure on the annotation of the pathological cases. However, gathering large datasets of normative populations is another challenge, since the vast majority of available clinical images are patient data with pathological patterns. Data augmentation techniques have been proposed as a way to address this issue in the data space [9,11,16]. One approach of data augmentation consists of explicit generation of synthetic instance based on synthesis or simulation methods as reviewed in [4]. Challenges in this domain have been addressed over the past few years through international events such as Sashimi workshops held in conjunction with the MICCAI conference. In this study we focus on synthesis methods based on deep neural architectures. The main trend has focused on the use of architectures based on segmentation networks such as U-Net, recently combined with adversarial branches as in generative adversarial networks (GANs) or Cycle-GAN for the synthesis of fake mono- or multi-modal modalities based on tuplets of paired (i.e. coregistered slices or volume of the same patient) or unpaired (i.e. paired of input and output training data that do not belong to the same patient or are not spatially coregistered) different modalities (e.g. multiple sequences of MRI, CT or PET) [16]. Research in this cross-modal translation or synthesis domain has been very productive over the past few years [2,8,10,13–15]. As far as we know, performance of most of the proposed architectures was evaluated based on visual quality metrics only, such as PSNR or turing test, but, only a few assessed task-based related performance, especially in the unsupervised anomaly detection context [15].
The objective of this study is to build on the recent MRI to PET GAN models to design an efficient architecture for the synthesis of realistic PET images derived from T1 MRI images of normal subjects. We do not only evaluate the visual quality of the synthetic PET data with standard metrics such as PSNR but also quantify their added value for the final medical task at hand. To that purpose, we consider the diagnostic task of epilepsy lesion detection in [^{18}F]fluorodeoxyglucose (FDG) PET exams. Following the idea proposed in [1], we build an automated unsupervised anomaly detection model that combines a feature extraction step based on a siamese network and a one-class SVM model. This model is trained on FDG PET brain exams of normal subjects. Our hypothesis is that increasing the size of the training dataset with synthetic FDG exams should translate into a gain in the epilepsy lesion detection rate. The main contributions of this paper are:

- A model derived from the Cycle-GAN architecture for the synthesis of realistic FDG PET exams of normal subjects from T1 MRI.
- A comparison of different variants of GANs methods based on the same training dataset.

– A global evaluation of the quality of these synthetic data including quantitative metrics of visual image quality as well as their ability to mimic real data for a diagnostic lesion detection task at hand.

2 Method

2.1 Synthesis of Realistic PET Data with GANs

In recent years, generative adversarial networks (GANs) [5] have demonstrated impressive results in computer vision and in biomedical image analysis, for sample generation, image synthesis, quality enhancement and image segmentation [14]. The basic structure of a GAN consists of the generator that is trained to generate new synthetic samples, and the discriminator that tries to distinguish examples being real or fake. These two models are trained simultaneously and compete against each other. In this study, we build on a comparative analysis of different variants of GAN architectures to design the optimal configuration with adapted loss terms for missing PET data generation from T1 MRI data.

GANs Architectures.
Simple-GAN. We first propose to use a standard GAN architecture with one generator G_B and one discriminator D_B (the upper part in Fig. 1). Generator G_B attempts to improve the quality of the translated output x_b of domain B from the original input y_A from the original domain A, thus deceiving the discriminator D_B. The training procedure is formulated as a min-max optimization problem of an objective function that the discriminator is trying to maximise and the generator is trying to minimize. In this study, we implement the least squares GAN (LSGAN) model [7] that aims to minimize the following discriminator $L_{LSGAN}(D_B, A, B)$ and generator $L_{LSGAN}(G_B, A, B)$ losses :

$$\begin{aligned} L_{LSGAN}(D_B, A, B) &= E_{p(x_b)}[D_B(x_b)^2] + E_{p(y_b)}[(D_B(y_b) - 1)^2] \\ L_{LSGAN}(G_B, A, B) &= E_{p(x_b)}[(D_B(x_b) - 1)^2] \end{aligned} \tag{1}$$

where y_a and y_b are true images of domain A and B, respectively, and $x_b = G_B(y_a)$ is the fake image of domain B generated from y_a.

In the context of supervised image translation, where the model can be trained on paired images in both domains at the pixel level (e.g. corresponding images of the same patient), we propose to add a mean squared error (MSE) loss term L_{mse} (see Eq. (2)) between the fake image x_b generated from a true image y_a of domain A and its paired true image y_b in domain B.

$$L_{mse}(G_B) = E_{p(x_b)}[(x_b - y_b)^2] \tag{2}$$

Cycle-GAN. Cycle-GAN consists of two generator networks G_A and G_B and two discriminator networks D_A and D_B. The baseline Cycle-GAN model is shown in Fig. 1. The generators translate images from domain A to domain B and vice

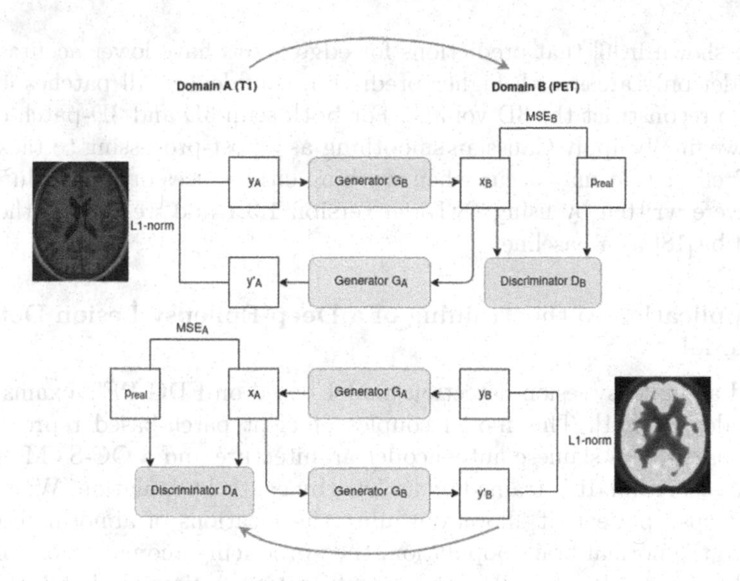

Fig. 1. Cycle-GAN architecture based on two baseline GANs translating images from domain A to domain B (upper GAN) and vice versa (lower GAN).

versa. Each of the generator networks is trained adversarially using a corresponding discriminator D_A and D_B. In addition to the adversarial loss term of the simple GAN network in Eq. (1), the key element in training Cycle-GAN network is a cycle-consistency loss function L_{cyc}:

$$L_{cyc}(G_A, G_B) = E_{p(y_a)}[\| y'_a - y_a \|_1] + E_{p(y_b)}[\| y'_b - y_b \|_1] \tag{3}$$

where y'_a is the fake image of domain A generated by generator G_A from the fake x_b, that is $y'_a = G_A(x_b)$ with $x_b = G_B(y_a)$. As for the simple GAN formulation, in a paired mode, we add a MSE loss term between real and synthetic images of both domains A and B.

Implementation Details
We consider two configurations depending on the size of the input data:

- ***semi-3D*** models which receive three adjacent transverse slices as input (each slice corresponding to one channel)
- ***3D-patch*** models where we feed 3D mini patches extracted from the original 3D images into the network

We take ResNet as the backbone architecture of both generators with 9 residual blocks for the semi-3D approach and 2 blocks for the fully 3D-patch configuration. PatchGAN is selected for the discriminators following the architectures proposed in [18]. In the semi-3D configuration, the whole 3D image is reconstructed by stacking the generated transverse slices. In the 3D-patch setting, we crop the generated 3D patches so as to consider only their central part as it has been shown in [6] that predictions for edge pixels have lower accuracy, thus we consider only areas with higher prediction confidence. All patches are then stacked to reconstruct the 3D volume. For both semi-3D and 3D-patch configurations, we finally apply Gaussian smoothing as a post-processing to tackle with "border" effect that may occur when stacking either slices or mini-volumes. All models were written by using PyTorch version 1.3.1 and we took python code provided by [18] as a baseline.

2.2 Application to the Training of a Deep Epilepsy Lesion Detection Model

We build an epilepsy lesion detection model based on FDG PET exams following the idea from [1]. This model couples efficient patch-based representation learning based on a siamese autoencoder architecture and a OC-SVM anomaly detection algorithm. It is trained on an healthy control population. When tested on an epilepsy patient, it allows outlining the locations of abnormalities with regards to the normal brain population, thus producing anomaly score maps. In this work, we reproduce the 2D siamese architecture depicted in Fig. 5 of [1] with 15×15 patch size, resulting in a feature vector of dimension 64 and perform a 4-fold cross validation to extract the latent variables of the control population. Images are scaled between 0 and 1 at image level before feeding the patches to the deep autoencoder architecture. We then build one oc-SVM model per voxel with RBF kernel in the latent representation space learned by the siamese autoencoder.

3 Experiments and Results

3.1 Data

This study including three clinical image databases was approved by our institutional review board with approval numbers 2012-A00516-37 and 2014-019 B and a written consent was obtained for all participants. The first database is used to compare the different deep generative models of synthetic PET data. It consists of a series of 35 paired FDG PET and T1 weighted MRI scans co-registered to the MNI space, thus leading to 3D image volumes of size 157×189×136 with 1 mm^3 isotropic voxel size. These data were acquired on 35 healthy volunteers on a 1.5T Sonata scanner and mCT PET scanner (Siemens Healthcare, Erlangen, Germany). The second database consists of 40 T1-weighted MRI exams acquired on healthy control subjects on the same 1.5T MR Sonata scanner. It is used to generate synthetic FDG PET data that then serve to train the brain anomaly detection model introduced in Sect. 2.2. The third database consists of 17 paired FDG PET and T1 weighted MRI scans of patients with confirmed medically intractable and subtle epileptogenic lesions, as illustrated in Fig. 3. These data were acquired on on the same 1.5T Sonata MR scanner. All control and patient PET scans are rigidly aligned to their corresponding MRI then co-registered to the MNI space with SPM12 (https://www.fil.ion.ucl.ac.uk/spm/doc/manual.pdf). All original MRI and PET 3D data are scaled between 0 and 1.

3.2 Generation of Synthetic PET Images from T1 Healthy Controls

For the semi-3D approach, we explore in total 4 variants of GANs for paired examples based on Simple-GAN or traditional Cycle-GAN architectures both with and without MSE loss. Forty-six triplets of adjacent slices per patient are extracted thus resulting in around 1 200 training samples for each model. For the 3D-patch based approach, we use Simple-GAN and Cycle-GAN both with additional MSE loss. 6 069 mini-patches of size $32 \times 32 \times 32$ are extracted for each healthy subject with a stride of 8, thus leading to more than 200 000 training mini-volumes. A 4-fold cross-validation performance study is conducted with 26 controls in the train set and 9 controls in the validation set. During the training, Structural Similarity Index (SSIM) [12] between real and synthetic validation images serves as a quality metric to define the optimal configuration (early stopping criterion). All semi-3D approaches and 3D-patch models are trained for a maximum of 200 and 100 epochs with a batch size of 5 and 10, respectively, and Adam optimizer. The learning rate of 0.0002 is kept constant for the 3D-patch models, while for the semi-3D models it is kept constant up to 100 epochs and linearly decayed to zero over the next 100 epochs. A 3D Gaussian smoothing is applied on the reconstructed PET images of all model types (semi-3D and 3D-patch) to reduce border effects. Among a range of values between 0 and 3 mm FHWM, the value of 1.5 mm is shown to produce the best SSIM values.

Table 1 reports the mean SSIM, Peak Signal to Noise Ratio (PSNR) and Learned Perceptual Image Patch Similarity (LPIPS) [17] with corresponding standard deviation computed over all validation samples and all folds for each of the six considered models. Semi-3D Cycle-GAN with MSE loss is shown to perform the best among the 4 semi-3D models considered in this study. Two-tailed Wilcoxon signed rank tests yield significant differences between the semi-3D and 3D-patch Cycle-GAN models with MSE loss for the PSNR (p-value $< 10^{-6}$) and LPIPS (p-value $< 2 \times 10^{-4}$) metrics. A p-value of 0.069 is also achieved for the SSIM metric. Also note that our proposition to add the MSE loss term to the Cycle-GAN global loss allows a significant improvement of all three metrics.

Table 1. Average visual quality metrics computed on the 35 synthetic PET exams generated from T1 MRI of 35 healthy subjects.

Configuration	Model	SSIM	PSNR	LPIPS
Semi-3D	Simple-GAN	0.818 ± 0.021	19.655 ± 1.441	0.035 ± 0.008
	Simple-GAN with MSE loss	0.879 ± 0.021	23.177 ± 1.760	0.022 ± 0.005
	Cycle-GAN	0.837 ± 0.028	21.750 ± 1.142	0.030 ± 0.006
	Cycle-GAN with MSE loss	$\mathbf{0.883 \pm 0.022}$	$\mathbf{23.525 \pm 1.388}$	$\mathbf{0.022 \pm 0.005}$
3D-patch	Simple-GAN with MSE loss	0.869 ± 0.031	19.852 ± 0.597	0.034 ± 0.017
	Cycle-GAN with MSE loss	0.875 ± 0.023	17.760 ± 1.767	0.026 ± 0.007

In the following, we consider the best performing of each configuration, namely semi-3D and 3D-patch Cycle-GAN models with MSE loss. Example synthetic PET data generated by these two configurations of Cycle-GAN models

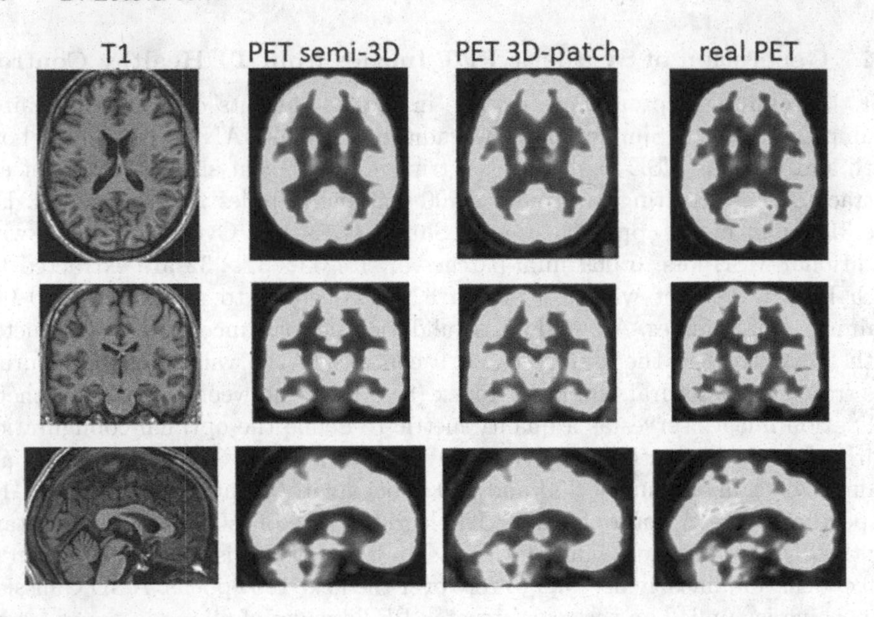

Fig. 2. Qualitative result on one control subject. Left column: original T1 MRI, right column: original PET image, the two central columns from left to right: synthetic PET image generated with semi-3D and 3D-patch Cycle-GAN models with MSE loss.

from the same T1 MRI of a control subject are illustrated in Fig. 2 and compared with the reference PET image of this subject. Both models allow generating visually realistic FDG PET data.

3.3 Application of Synthetic PET Data to the Training of a Brain Anomaly Detection Model for Epilepsy Patients Screening

Our second objective is to demonstrate that the realistic synthetic PET data can serve to improve performance of machine learning based diagnostic models. The considered application is the brain anomaly detection model described in Sect. 2.2. This model is trained on three different databases: the series of 35 real control PET dataset described in Sect. 3.1 and two hybrid databases mixing these 35 real control PET with 40 synthetic control FDG PET data generated by the semi-3D and 3D-patch Cycle-GAN models (with MSE loss), respectively. These 3 models are then tested on 17 patients with confirmed medically intractable epileptogenic lesions. Note that these patients correspond to difficult detection cases. Their FDG PET exam is indeed considered as normal, meaning that the hypometabolic lesions are subtle and barely visible by naked eye. Results reported in Fig. 4 indicate that the best detection sensitivity of 64.7% was achieved with the model trained on the hybrid dataset including the 40 PET data generated from the best semi-3D model. Adding these synthetic data to the training, which here amounts to doubling the number of training samples, allows a 20% gain in sensitivity compared with that achieved with the

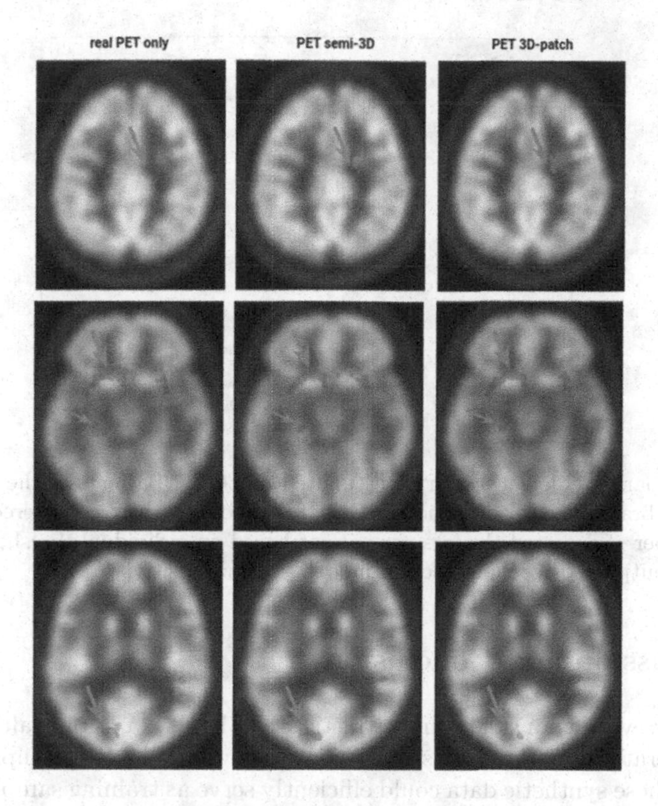

Fig. 3. Example cluster maps for three patients produced by the detection models, from left to right: 35 real PET scans, 35 real+40 synthetic PET (semi-3D Cycle-GAN), 35 real+40 synthetic PET (3D-patch Cycle-GAN). Blue arrows point to suspicious anatomical regions. The upper line demonstrates a case where both models trained with additional synthetic data managed to detect a lesion with a high confidence (bright green color) in the right internal frontal lobe, while it is missed by the model trained on real PET data solely. The middle line shows a successful case, where all models detect two clusters (green coloured) in the left internal temporal lobe and hippocampus. For the bottom line patient, models trained on original PET and with synthetic PET (semi-3D) managed to detect a lesion in the left lateral remainder of occiptal lobe, but the correct location is missed for the 3D-patch Cycle-GAN model. Red clusters correspond to false positives (the brighter the color the higher the rank of the cluster). (Color figure online)

same model trained on 35 real PET scans only achieving a 53% sensitivity. Performance achieved with the hybrid dataset including 40 PET data generated from the 3D-patch model reached a 41% sensitivity which is lower than that achieved with the model trained on the 35 real PET exams. Figure 3 illustrates anomaly maps derived from the three detection models on three test epilepsy patients.

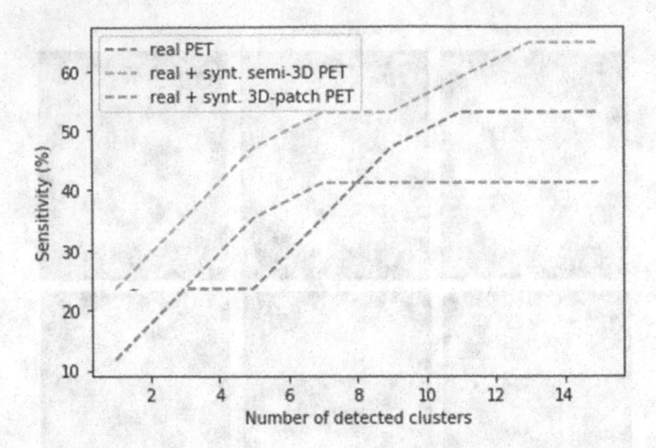

Fig. 4. Comparative detection curves estimated on the 17 patients of the test dataset based on the brain anomaly detection models trained on the three considered databases. x-axis: number of detected clusters per patient based on individual thresholding of the score maps outputted by the detection model, y-axis: sensitivity.

4 Discussion and Conclusion

In this study, we demonstrate that realistic FDG PET exams of healthy subjects can be generated from GAN based architectures with T1 MRI as input. We also show that these synthetic data could efficiently serve as training samples to boost the performance of machine learning based diagnostic models.

As seen in Fig. 2, both semi-3D and 3D-patch Cycle-GAN models produce PET images which closely match the original ones. In both cases, however, the histogram of the intensity does not perfectly match that of the original data. One perspective regarding this issue would be to further constrain the generative model to match the global intensity value of the true PET image or perform histogram matching. This may positively impact the performance of the brain anomaly detection model. Figure 4 shows that performance achieved with the added synthetic 3D-patch data are lower than that achieved with the true PET training dataset. Paired visual analysis of the anomaly maps generated by models trained with semi-3D and 3D-patch synthetic PET indicate very similar patterns for all 17 patients, except for two of them which did not contain any suspicious cluster in the lesion anatomical region for 3D-patch unlike in the produced anomaly map based on the semi-3D approach. Further analysis is required to better understand this observed difference. We also plan to add more patient to this analysis to evaluate if the trend observed in Fig. 4 is confirmed.

The best performing model allows achieving a detection sensitivity of 64% which may seem low. Note that this value has to be compared with very low sensitivity of human experts on these difficult diagnostic cases and is in par with reported values in a recent study questioning the added value of synthetic PET data for the same clinical application [15].

Acknowledgement. This work received funds from the Région Auvergne-Rhône-Alpes through the TADALOT project. It was performed using HPC resources from GENCI-IDRIS (Grant 2021-AD011011938) and within the framework of the LABEX PRIMES (ANR-11-LABX-0063) of Université de Lyon operated by the French National Research Agency (ANR).

References

1. Alaverdyan, Z., Jung, J., Bouet, R., Lartizien, C.: Regularized siamese neural network for unsupervised outlier detection on brain multiparametric magnetic resonance imaging: application to epilepsy lesion screening. Med. Image Anal. **60**, 101618 (2020)
2. Armanious, K., Jiang, C., Abdulatif, S., Küstner, T., Gatidis, S., Yang, B.: Unsupervised medical image translation using cycle-medgan. In: 2019 27th European Signal Processing Conference (EUSIPCO), pp. 1–5. IEEE (2019)
3. Baur, C., Denner, S., Wiestler, B., Albarqouni, S., Navab, N.: Autoencoders for Unsupervised Anomaly Segmentation in Brain MR Images: A Comparative Study. arXiv:2004.03271 [cs, eess] (2020)
4. Frangi, A.F., Tsaftaris, S.A., Prince, J.L.: Simulation and synthesis in medical imaging. IEEE Trans. Med. Imaging **37**(3), 673–679 (2018)
5. Goodfellow, I., et al.: Generative adversarial nets. In: Advances in neural information processing systems, pp. 2672–2680 (2014)
6. Huang, B., Reichman, D., Collins, L.M., Bradbury, K., Malof, J.M.: Tiling and stitching segmentation output for remote sensing: Basic challenges and recommendations. arXiv preprint arXiv:1805.12219 (2018)
7. Mao, X., Li, Q., Xie, H., Lau, R.Y.K., Wang, Z., Smolley, S.P.: Least squares generative adversarial networks. In: Proceedings of the IEEE Conference on Computer Vision (ICCV), pp. 2813–2821 (2017)
8. Pan, Y., Liu, M., Lian, C., Zhou, T., Xia, Y., Shen, D.: Synthesizing missing pet from MRI with cycle-consistent generative adversarial networks for Alzheimer's disease diagnosis. In: Frangi, A.F., Schnabel, J.A., Davatzikos, C., Alberola-López, C., Fichtinger, G. (eds.) Medical Image Computing and Computer Assisted Intervention - MICCAI 2018, pp. 455–463. Springer International Publishing, Cham (2018)
9. Shorten, C., Khoshgoftaar, T.M.: A survey on image data augmentation for deep learning. J. Big Data **6**(1), 60 (2019)
10. Sikka, A., Peri, S.V., Bathula, D.R.: MRI to FDG-PET: cross-modal synthesis using 3D U-net for multi-modal Alzheimer's classification. In: Gooya, A., Goksel, O., Oguz, I., Burgos, N. (eds.) SASHIMI 2018. LNCS, vol. 11037, pp. 80–89. Springer, Cham (2018). https://doi.org/10.1007/978-3-030-00536-8_9
11. Tajbakhsh, N., Jeyaseelan, L., Li, Q., Chiang, J.N., Wu, Z., Ding, X.: Embracing imperfect datasets: a review of deep learning solutions for medical image segmentation. Med. Image Anal. **63**, 101693 (2020)
12. Wang, Z., Bovik, A.C., Sheikh, H.R., Simoncelli, E.P.: Image quality assessment: from error visibility to structural similarity. IEEE Trans. Image Process. **13**(4), 600–612 (2004)
13. Wei, W., et al.: Predicting pet-derived demyelination from multimodal MRI using sketcher-refiner adversarial training for multiple sclerosis. Med. Image Anal. **58**, 101546 (2019)

14. Wolterink, J.M., Leiner, T., Viergever, M.A., Išgum, I.: Generative adversarial networks for noise reduction in low-dose ct. IEEE Trans. Med. Imaging **36**(12), 2536–2545 (2017)
15. Yaakub, Siti Nurbaya, McGinnity, Colm J.., Clough, J.R., Kerfoot, E., Girard, N., Guedj, E., Hammers, A.: Pseudo-normal PET synthesis with generative adversarial networks for localising hypometabolism in epilepsies. In: Burgos, N., Gooya, A., Svoboda, D. (eds.) SASHIMI 2019. LNCS, vol. 11827, pp. 42–51. Springer, Cham (2019). https://doi.org/10.1007/978-3-030-32778-1_5
16. Yi, X., Walia, E., Babyn, P.: Generative adversarial network in medical imaging: a review. Med. Image Anal. **58**, 101552 (2019)
17. Zhang, R., Isola, P., Efros, A.A., Shechtman, E., Wang, O.: The unreasonable effectiveness of deep features as a perceptual metric. In: Proceedings of the IEEE Conference on Computer Vision and Pattern Recognition, pp. 586–595 (2018)
18. Zhu, J.Y., Park, T., Isola, P., Efros, A.A.: Unpaired image-to-image translation using cycle-consistent adversarial networks. In: Computer Vision (ICCV), 2017 IEEE International Conference on (2017)

Correction to: Frozen-to-Paraffin: Categorization of Histological Frozen Sections by the Aid of Paraffin Sections and Generative Adversarial Networks

Michael Gadermayr, Maximilian Tschuchnig, Lea Maria Stangassinger, Christina Kreutzer, Sebastien Couillard-Despres, Gertie Janneke Oostingh, and Anton Hittmair

Correction to: Chapter "Frozen-to-Paraffin: Categorization of Histological Frozen Sections by the Aid of Paraffin Sections and Generative Adversarial Networks" in: D. Svoboda et al. (Eds.): *Simulation and Synthesis in Medical Imaging*, LNCS 12965, https://doi.org/10.1007/978-3-030-87592-3_10

In an older version of this paper, there was an error in the affiliation of the author Sebastien Couillard-Despres. This has been corrected.

The updated version of this chapter can be found at
https://doi.org/10.1007/978-3-030-87592-3_10

D. Svoboda et al. (Eds.): SASHIMI 2021, LNCS 12965, p. C1, 2022.
https://doi.org/10.1007/978-3-030-87592-3_15

Author Index

Printed in the United States
by Baker & Taylor Publisher Services

Printed in the United States
by Baker & Taylor Publisher Services